Teaching Algebra in the Middle Grades:

A Professional Development Guide

by Joseph C. Power

HOLT, RINEHART AND WINSTON

A Harcourt Classroom Education Company

Austin · New York · Orlando · Atlanta · San Francisco · Boston · Dallas · Toronto · London

Dear fellow teacher,

Perhaps you are called upon to teach algebra for the first time. If so, you may be curious about finding resources that can help you understand what algebra is about, where its ideas come from, and how its concepts and skills are related. Perhaps you are also asking yourself where you can find assistance in teaching algebra to students who may have limited experience in the subject.

My hope is that the lessons in this book will provide answers to your pedagogical queries as well as ideas about how to teach algebra for understanding. As you read this book, you will see that it covers the significant topics found in the student textbook. Upon visiting various lessons, you will be able to discern the book's emphasis on procedural knowledge (the "how-to" aspect of algebra) and conceptual knowledge (the "why" aspect of algebra). For example, you will notice many questions are posed and then thoughtfully answered. This approach may help you think through various topics with the guidance of an experienced mathematics teacher.

You will also notice various whiteboard simulations. They are intended to reveal how one mathematical thinker can reason thoughtfully through a problem. Another feature that you may find useful is the selection of problems similar to those found in the student textbook but presented in a way that shows how the solution steps are linked. Arrows are used to illustrate the links. Different fonts are used to emphasize explanations and describe thinking processes.

Writing this book has given me the opportunity to share what I know about mathematics, its roots, its teaching, and mathematical thinking patterns. Naturally, I had to deliberate about how to create a book that teachers can use as their professional development guide. Eventually I decided to frame the discussions in this book somewhat like conversations with you and your colleagues.

I have written this book so that you can pick and choose to read the parts of it that you most wish to study. In this way, the book becomes a practical and flexible resource. Just as students develop mathematically through mathematical experience, you can develop mathematically with the aid of this professional development book.

Best wishes in your professional development and teaching career.

Joseph C. Power

1 The Role of the Mathematics Teacher in the Middle Grades

2 Linear Equations and Inequalities in One Variable

3 Linear Equations and Inequalities in Two Variables

4 Systems of Linear Equations and Inequalities

5 Polynomials and Factoring

6 Quadratic Equations and Functions

7 Rational Expressions and Functions

8 Functions

9 Mathematical Structure and Reasoning

THE ROLE OF THE MATHEMATICS TEACHER IN THE MIDDLE GRADES

The Task

At all grade levels, the mathematics teacher is a member of an educational team whose common purpose is intellectual development in a variety of subject areas. These subject areas include social studies, geography, science, language arts, and mathematics. While the scope of content is not limited to these disciplines, those mentioned occupy a high percent of the team's attention. Where does the mathematics teacher fit in this system? What specific objectives is the mathematics teacher in the middle grades expected to achieve?

In general, there are two content agendas for the middle grades mathematics teacher. These are numerical work and algebraic work. In the middle grades, great emphasis is placed on the the teaching of mathematical language, properties, and methods involving variables as well as numerical work with whole numbers, fractions, decimals, and percents. During the progression through middle grades, the teaching of algebraic skills increases as does the expectation that numerical skills have been learned. This thought is roughly illustrated in the diagram below.

Grade 6	numerical work	algebraic work
Grade 7	numerical work	algebraic work
Grade 8	numerical work	algebraic work
Beyond	all algebraic work	

This illustration suggests a gradual transition from major preoccupation with arithmetic skills to a greater concern for algebraic work relying on numerical skills. The illustration is not meant to suggest any percents according to which time should be allotted to the various agendas.

As children grow intellectually, their ability to think in abstract terms develops. The illustration above suggests that as students move through the grades, they are expected to understand a greater number of abstract concepts and general skills involved with algebra. Abstraction allows students to add, subtract, multiply, and divide letters as they would do with numbers. As students become abstract thinkers, their ability to manipulate symbolic algebraic structures as if they were physical objects increases. In particular, students move from seeing the equations below as all different from one another toward seeing that, while they appear to be different from one another, they are all solved by the same method.

$$x + 3 = 7 \qquad 3 + t = 7 \qquad 7 = 3 + w \qquad 7 = a + 3$$

Subtraction Property of Equality:
Subtracting the same amount from two quantities that are equal to one another preserves their equality.

All of these equations are solved by application of the **Subtraction Property of Equality.** As attention paid to algebraic work increases, students learn to make the transition from verbal to symbolic statements as shown below.

Verbal: If a quantity is taken from two equal quantities, the resulting quantities are equal.

Symbolic: If a, b, and c are real numbers and $a = b$, then $a - c = b - c$.

With this general overview in mind, we can categorize mathematics learning as procedures (the "how-to" aspect of mathematics) and concepts (the "why" aspect of mathematics).

Mathematics as Procedures (Procedural Knowledge)

procedural knowledge:
The ability to follow a predetermined set of procedures. Performing procedures without a conceptual understanding is meaningless learning.

Procedural knowledge is the type of knowledge that allows students to perform operations, apply formulas, use measurements, and calculate results. Procedural knowledge provides quantitative results that students should interpret in light of the meaning, logic, and reason supporting these procedures.

Calculation

- Teach students to add, subtract, multiply, and divide whole numbers.

- Teach students to add, subtract, multiply, and divide positive decimals.

- Teach students to add, subtract, multiply, and divide fractions and mixed numbers.

- Teach students to use relationships between numbers such as comparing and ordering different numbers.

- Teach students to find and use percent.

- Teach students to perform operations on positive and negative numbers.

Measurement

- Teach students how to use measurement tools, such as a ruler and protractor.

- Teach students to use geometry formulas to find length, area, and volume.

- Teach students to deal with units of measure in different measurement systems, in particular the English and metric systems of measurement.

- Teach students to use calculation skills to solve problems in contexts such as applications and geometry.

Geometry

- Teach students how to recognize geometric shapes according to essential characteristics and definitions.

- Teach students to recognize, write, and use relationships among geometric objects.

Data

- Teach students how to find basic statistical measures such as mean, median, and mode.

- Teach students how to represent numerical data in graphs and charts such as bar graphs and circle graphs.

Algorithms

- Teach students the procedures by which correct calculations are obtained.

- Teach students procedures by which straightforward problems are solved.

algorithm: A step-by-step procedure for solving a specific kind of problem.

This categorization with its attendant lists can be expanded. Indeed, curriculum guidelines or standards outline in detail skills and content areas that must be covered at each grade level to ensure that the student is prepared to show competence in mathematics at that specific grade level.

Teaching algebra skills in a sequential order does not mean that they should be taught as independent from one another. Rarely do algebra problems require the application of one single skill. Instead, multiple skills are necessary to solve a problem. For example:

- Students need to know how to add decimals in order to find the mean of a set of decimals.

- Students need to know how to add, subtract, multiply, and divide whole numbers, fractions, and decimals in order to use a geometry formula to find area.

Because of this, teachers in individual grade levels frequently teach more than one educational objective at a time. Just how much and how far the teacher goes in one grade level depends on the scope and sequence of the curriculum.

The organizational breakdown on the preceding pages is based on the rather mechanical part of mathematics, namely skills and procedures. Another categorization of the mathematical task is based on concept formation and analytical skills. It should be noted that these categorizations are not contradictory or mutually exclusive. In fact, successful mathematics learning implies an ability to connect in a meaningful way the procedural knowledge with the conceptual knowledge.

Mathematics as Concepts (Conceptual Knowledge)

conceptual knowledge: The ability to convert concrete to abstract and vice versa. A concept is an abstract generalization, category, or definition with a variety of concrete examples. One goal of mathematics instruction is that students recall a concept that is associated with a specific example.

Algebra and Functions

- Teach students how to apply mathematical properties of addition, multiplication, and number to solve equations and inequalities whose difficulty level is determined by grade level.

- Teach students to represent a verbal problem as a mathematical equation or inequality, solve the equation or inequality, and answer the question embedded in the problem.

- Teach students to make and use functional relationships between quantities represented by variables.

Strategies, Including Problem-Solving Strategies

- Teach students to read a problem carefully, noting what information is given, what is unknown, and how known information and procedures can deliver a solution for the unknown.

- Teach students to choose methods appropriate to a given problem, for example, what operation to choose, how to break a complicated problem into simpler problems.

- Teach students to use specific problem-solving strategies, such as drawing a diagram or rewriting a problem in simpler form as aids to solving a problem whose solution method is not immediately evident.

Abstraction

- Teach students to recognize patterns, in particular, to recognize that two apparently different problems are essentially the same from the point of view of solution.

- Teach students to read the symbolic language used in mathematics books.

- Teach students to write symbolic language to communicate their mathematical ideas.

- Teach students to understand that procedures and rules are generalizations and have universal applicability within their appropriate domains.

- Teach students to translate nonmathematical language into mathematical language and use mathematical generalizations to arrive at calculations, conclusions, or other results.

- Teach students that many real-world objects have mathematical representations. For example, a circle is a mathematical object created by looking at objects such as wheels in real life.

Reasoning

- Teach students to use deductive reasoning in various contexts. In particular, teach students that logical completion of an algorithm is deductive reasoning that gives conclusions with certainty.

- Teach students to chain relationships or mathematical skills in a logical way to arrive at conclusions, answers, or solutions.

- Teach students to give mathematical reasons for mathematical steps taken.

- Teach students to examine multiple instances in a situation and draw reasonable inferences about patterns perceived.

Number Sense

- Teach students how to compare numbers according to size and form, and how seemingly different forms can represent equivalent values.

- Teach students to use estimation skills to solve problems or serve as an aid to solving problems.

An important implication drawn from these two categorizations is that algebra procedures should not be taken as the end of the mathematical task but rather as a means to reach important mathematical concepts.

Unfortunately, it is not uncommon that students focus all their attention on memorizing procedures only. As a result, they become rule-based learners as opposed to concept-based learners. A rule-based learner knows how to carry out procedures but has limited understanding of why he or she is doing this work.

Concept-based learners, on the other hand, use procedures as a vehicle for their logical mathematical thinking. An example involving distance and mixed numbers will illustrate how concept-based students learn and apply both procedures and logic in order to solve a problem. The problem is shown below.

Problem

Helen and David begin bicycling toward one another on a straight road at the same time. Initially, they are 30 miles from one another. In 1 hour, Helen bicycles $12\frac{3}{8}$ miles and David bicycles 14.4 miles. How far apart are they after 1 hour?

Understanding the Problem

Read for given information carefully. Represent the verbal problem graphically by drawing a diagram.

The arrows pointing toward one another represent distance. The unknown distance between Helen and David is the distance between the arrowheads.

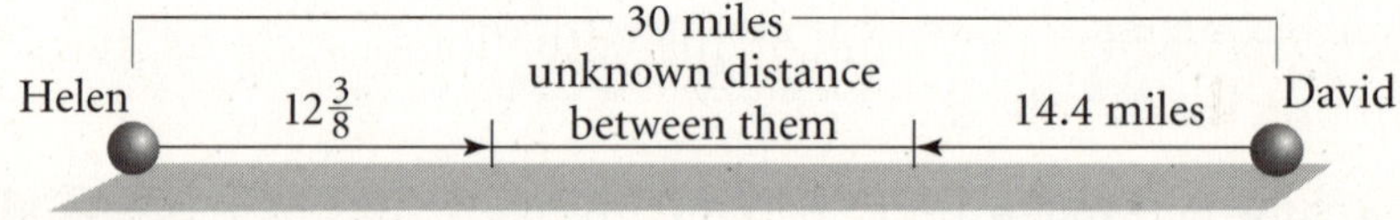

Arithmetic Solution

Reasoning: Relate distances. The unknown distance between Helen and David is 30 miles less the distances they each traveled.

Alternative reasoning: The sum of Helen's distance, David's distance, and the distance between them is 30 miles.

Strategy: Add distances traveled and subtract from 30.

Algorithm: Change 14.4 to a mixed number. $14.4 = 14\frac{2}{5}$

Algorithm: Add mixed numbers. $12\frac{3}{8} + 14\frac{2}{5} = 26\frac{31}{40}$

Algorithm: Subtract a mixed number $30 - 26\frac{31}{40} = 3\frac{9}{40}$
from a whole number

Number sense: The distance they travel is about 26 miles or 27 miles. It makes sense they would be more than 3 miles but less than 4 miles apart after 1 hour.

The solution above involves some of the processes and concepts discussed. Each process and conceptual understanding can be interpreted as smaller tasks into which the problem solver has divided the bigger problem. Changing a decimal into a mixed number, adding two mixed numbers, and subtracting a mixed number from a whole number are each simpler problems. Solving each small problem contributes to solving the given problem. Transforming a problem into smaller, more manageable tasks is a problem-solving strategy that students should learn as they practice.

Part of the role of the mathematics teacher in the middle grades is to help students integrate procedural and conceptual knowledge for the purpose of solving problems that are greater than the sum of their parts.

As the teacher succeeds in his or her task, students might be expected to write a solution to the problem above in a faster and more efficient way. For example, students may read such a problem, quickly jot down an expression such as $30 - \left(12\frac{3}{8} + 14\frac{2}{5}\right)$, carry out the arithmetic, and report the answer. Students who mentally work through strategies and reasoning should be prepared to articulate the reasoning that underlies their written work.

Students who study algebra as part of their middle grades mathematics education would be expected to use mathematical properties that are applicable in many problem-solving situations, such as the distance problem previously solved.

- The Subtraction Property of Equality is used as part of the algebraic solution to the distance problem.

- An important fact from geometry is illustrated below. This fact is implicitly used in both the arithmetic and algebraic solutions to the distance problem.

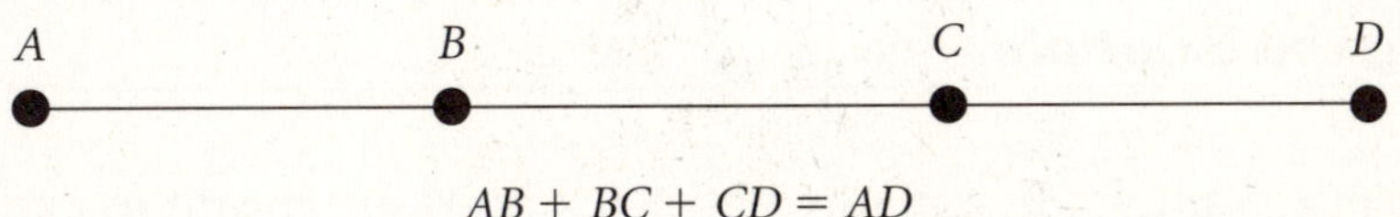

$$AB + BC + CD = AD$$

Algebraic Solution

Reasoning: Relate distances. The unknown distance between Helen and David is 30 miles less the distances they each traveled.

Alternative reasoning: The sum of Helen's distance, David's distance, and the distance between them is 30 miles.

Strategy: Write and solve an equation. Let x represent the distance between Helen and David.

Abstraction: $12\frac{3}{8} + x + 14.4 = 30$

$$12\frac{3}{8} + x = 30 - 14.4 \qquad \textit{Subtraction Property of Equality}$$

$$12\frac{3}{8} + x = 15.6 \qquad \textit{Subtraction}$$

$$x = 15.6 - 12\frac{3}{8} \qquad \textit{Subtraction Property of Equality}$$

$$x = 3\frac{9}{40} \qquad \textit{Subtraction}$$

Number sense: The distance they travel is about 26 miles. It makes sense they would be less than 4 miles apart after 1 hour.

By using a numerical or algebraic approach, the students obtain the same solution.

One of the differences between a student who uses arithmetic and a student who uses algebra is the fact that the algebra student has learned how to apply general mathematical properties to solve problems. Thus, the algebra student takes a step forward toward integrating skills and strategies in the problem-solving process. Notice that an algebra student cannot be a successful problem solver without knowledge of arithmetic skills.

Accomplishing the Task

How does the middle grades mathematics teacher accomplish the task of helping students become mathematically competent?

Setting the Example

Worked-out examples in a mathematics textbook are meant to explain and illustrate what the students should do in order to succeed. The same can be said of a whiteboard demonstration. Obviously, it is not enough to tell students that they need to learn how to perform operations with numbers. Students must be shown with examples how to do their mathematical work.

Teachers need to exemplify the thinking processes that students are to adopt and develop. When the teacher thinks out loud, writes sentences or equations on the whiteboard, or sketches diagrams, the teacher is modeling important learning behaviors for students.

The teacher's positive attitude toward mathematical knowledge and thinking processes is also a key element in setting the example. Students develop likes and dislikes partly because of likes and dislikes perceived in others. Teachers who praise students for their successes help students develop a positive attitude toward mathematics.

Practice Opportunities

The more students practice a particular skill, the more they will become adept in applying it. Students need to practice a skill immediately after the skill is taught. Follow-up exercises in textbooks provide this type of practice. Exercise sets provide additional exercises that the student can work in class, in a study group, or at home.

Students should be encouraged to keep a journal in which they can record daily progress. In particular, students should be encouraged to jot down the type of problems that cause them difficulty so that they can ask for assistance when returning to class the next day.

Some distinctions between meaningful vs. mechanical practice deserve attention here.

- Students do not learn algebra skills by volume. Students may well work many practice problems and get them all wrong for some reason. Monitoring student practice is necessary. If practice is not enough, then alternative teaching approaches may be necessary.

- Practice by itself does not provide conceptual understanding. Students may practice many problems, get correct answers, and still not understand how it all works. Put another way, students may develop good procedural skills without automatically developing logical and conceptual skills.

In short, learning is an active process that requires monitored practice, reflection, analysis, and self-evaluation.

Human Interaction

Students can learn a great deal from fellow students. For example, when students check one another's answers, take turns at solving problems that require multiple steps, verbalize concepts and ideas in their own words, and solve problems on the whiteboard, they are sharing mathematical ideas with the other members of the learning community, thus building mathematical knowledge for all. Likewise student-teacher interactions contribute to build meaningful knowledge. As a student works through a problem, the teacher can observe and assess the depth of a student's thinking processes, assist in moving these processes to deeper levels, and pose guiding questions that help the student get started on learning strategies.

Interaction with a teacher can also help generate a sense of self-importance. Students who feel that their work and effort is valued by the teacher usually develop high levels of motivation and self-esteem.

metacognitive skills: Cognitive skills that involve knowing how to organize one's knowledge, how to learn, how to solve problems, and how to correct one's errors in understanding.

As mentioned earlier, one of the goals of the mathematics teacher is to help the student become an independent thinker. Meaningful teacher-student interactions allow the teacher to perceive whether or not the student is developing **metacognitive skills,** or the ability to control their own learning. Test results, whether from a standardized test or one created by the teacher, may not be adequate instruments to assess a student's intellectual maturity.

Teacher-student interaction provides the teacher with a wealth of knowledge about students.

- Through work with individual students, teachers can get a sense of whether a particular type of teaching approach is working. Interaction is perhaps the best way to create a better match between learning styles and teaching styles.

- Teachers who listen carefully to what students say will gain insight into how students think mathematically and how secure they feel in the mathematical work they do. This can have an effect on how the teacher plans lessons, assesses students, and adapts curriculum materials to meet the students' needs.

- Lastly, teachers can learn from students' alternative explanations to problems. Both the teacher and the students belong to a community of learners in which every member practices the art of explanation. For example, consider the two different presentation styles below for explaining how to solve $a + 5.5 = 7.9$.

$$a + 5.5 = 7.9 \qquad\qquad a + 5.5 = 7.9$$
$$a + 5.5 - 5.5 = 7.9 - 5.5 \qquad\qquad -5.5 = -5.5$$
$$a = 2.4 \qquad\qquad a = 2.4$$

Some teachers may find through interaction that a particular explanation style may work better with some students.

Textbooks and worksheets do not smile, speak encouraging words, or engender feelings of security and hope. Nor does a calculator or computer screen. Interaction with others, particularly the teacher, is perhaps the one ingredient that cannot be replaced by the printing press or the computer industry. Many students in later life comment that it was the teacher that made the difference. A positive and memorable experience with a mathematics teacher usually results in a positive attitude toward mathematics as a discipline.

Much has been said about student-student interaction and teacher-student interaction. There is, however, another interaction that is crucial for the teacher. This interaction is with fellow teachers. Younger teachers find advice and guidance from more experienced teachers helpful. In some cases, interaction may involve how to teach a particular topic. At other times, it may involve forms of reteaching or assessment or interacting with students.

Assessment

Both teachers and students need assurance that educational goals have been met. This assurance is provided by assessment, which should be understood in its broadest sense. Assessment encompasses traditional forms of testing as well as formation of judgment based on evidence other than observation of accuracy. An example will illustrate what is meant here.

$$2x + 5 = 17$$
$$2x = 12$$
$$x = 6$$

A problem such as the one at left appears on a multiple-choice test. A student jots down the problem on paper. The correct choice is C and the student circles C. A check of the student's test paper indicates a technical competence. In a personal interview with the student, the teacher asks the student to identify the algebraic properties used to support the answer chosen. The student stumbles and finally declares "I don't really know what you want. I simply did what I was shown in class." The teacher may reasonably infer that the student has not developed an appreciation that mathematical work is supported by articulation of mathematical reasoning and properties.

Applications of Mathematics

In mathematics, application problems serve various purposes.

- The application might be one that cannot be solved by using only what students learned previously. Thus, students have a reason to learn a new method or skill.

- The application situates a mathematics problem within a well-defined context, thus providing the student with information that they can relate to other information they already know.
- The application, if well chosen and clearly presented, should remind students that one of the reasons people create mathematics is to solve practical problems.

In other disciplines, such as physics or Earth science, applications form the core of the subject matter. In mathematics, applications are an important link between the highly symbolic structure of mathematics and the meaning that these symbols represent. Another benefit for studying mathematics with applications is the creation of multiple representations of problems, which contributes to enhance students' understanding of mathematics. Some students may approach an application problem by using a table or a graph, while others may prefer a formula or an equation.

Use of Actual Objects

Many mathematical concepts and properties have their roots in experience. Shown below are two collections of pencils. Notice that the number of pencils in Collection A is 3 less than the number of pencils in Collection B. A classroom setup like this can be used to help students understand important mathematical properties, such as the Property of Inequality explained below.

Collection A: 6 pencils

Collection B: 9 pencils

Remove 4 pencils.   Remove 4 pencils.

When 4 pencils are removed from each collection, Collection A will still have 3 fewer pencils than Collection B. Using actual objects can help students make the generalization below.

If $a < b$, subtracting c from a and from b will give a true inequality, $a - c < b - c$.

Manipulating actual objects allows students to reconfirm the truth of the mathematical property. Students understand abstract ideas better when those ideas are presented with connections to experiences in the real world. Object manipulation, however, should be used only as a means to teaching algebra, not as an end.

Discovery Activities

An activity in which students discover something by themselves can contribute to empowerment. In general, people take better possession of the knowledge they discover on their own than knowledge that is imparted by dictate. Discovery activities offer students a number of benefits. For example, students can seize opportunities to develop leadership skills. They can guide their own learning and develop metacognitive skills. They can also appreciate contributions made by other students to a discovery learning activity. Discovery activities work well when they take their place in the wide set of teaching options.

Cooperative Learning

In the mathematics classroom, a cooperative learning activity brings the otherwise contrived context of a classroom closer to the human interaction that characterizes our everyday activities. When students work with one another and with the teacher's guidance, mathematics ceases to be an abstract body of knowledge separated from reality. The activity gives both students and teacher a chance to do and learn mathematics purposefully. Cooperative learning activities also give students the opportunity to communicate in their own words the mathematical processes and ideas being used in the activity. Finally, cooperative learning activities provide students the opportunity to be actively engaged in their own learning. Passive learning tends to engender more problems than benefits as students tend to become bored with lengthy teacher lectures and, as a result learn less than active learners.

Whiteboard Demonstrations

This technique provides a way for students to see how the teacher would solve a particular problem. Students may or may not be actively involved in filling in the work at the whiteboard. Nonetheless, the demonstration serves as a model that students can imitate when beginning their first practice problem of the same sort. Many teachers find that the length of time spent on a particular demonstration should not be too long. Demonstrations that are too lengthy can cause students to lose their concentration, focus, and interest.

Worksheets

A worksheet cannot replace meaningful instruction. Instructions provided in a worksheet should not be confused with instruction provided by the teacher. The instructions on a worksheet are directives that students carry out.

- The directive may be as simple as "Solve each equation." Worksheets serve the purpose of providing practice opportunities.
- A worksheet may show a worked-out example, followed by a directive such as "Write the reason that justifies each step." The purpose of this part of the worksheet might be to help students develop reasoning skills. A second directive on the worksheet might read "Write a complete solution for the problem below. To the right of each step, write the reason justifying that step." The purpose of this part of the worksheet might be to help students articulate their reasoning and enhance their skills in writing mathematics.

Many textbooks are accompanied by worksheet booklets. Teachers should be careful when choosing the type of worksheet that is most appropriate for a lesson. To the extent that teachers feel comfortable creating their own worksheets, they can customize the learning experience even more.

Transparencies

A textbook example can be reproduced as a transparency for the purpose of explaining the process of solving a problem in an orderly way. To do this, the teacher may adapt a textbook example by adding explanatory side notes to a worked-out example. The teacher can then place the transparency on an overhead projector and lead a class discussion based on the problem. Some advantages to doing this include:

- Students are not looking down at the textbook page. Rather their attention is directed to a central point in the problem.
- The teacher need not repeat on the whiteboard what is already written in the book. The teacher is then free to explain each step taken, ask questions to students, and allow students to ask questions. By uncovering one part of the transparency at a time, students can see the step-by-step solution unfold.

Continuing Education

Many companies provide on-the-job training for their employees. Reasons for doing this include:

- To train new employees in jobs for which previous training is either insufficient or nonexistent.
- Even experienced employees may undertake on-the-job training because new technologies are introduced into the company, the company has redefined a part of its focus, or it has realigned its priorities.

- Changing from one job to a different but related job may require continuing education in the form of new on-the-job training.

On-the-job training may occur in various forms. It may take the form of a course of study, a workshop, a seminar, or mentoring. Employees may even travel to a different site for some period of time to learn skills not available at the employee's current job site. It should also be noted that some employees continue their education through informal rather than formal interaction with others. This type of continuing education takes place gradually on a daily basis and is not marked with a formal starting point or ending point.

In the educational community, the need for continuing education is no less evident or important. There are many reasons for this.

- The student population changes. As a result, what works with one population may or may not work as well with a different population or different segments of it.

- Teaching assignments change. Teachers who have taught at one particular grade level may find it necessary to investigate what adaptations are needed to adjust to a higher or lower grade level. Other teachers may take on a subject area that was not part of their formal training in college. These teachers need specialized training in order to teach the content of the new subject area successfully.

- Teachers who left their teaching careers and then return to teaching may find that the teaching field has changed. For example, new approaches to teaching and learning may have replaced older ones. The degree of student involvement may have escalated. Adherence to curriculum standards may be more rigorous than before. Indeed, the curriculum guidelines may be different than before. Whatever the reason, teachers in this group may experience some unexpected culture shock. As a result, these teachers may feel the need to bring themselves up to date in terms of what is happening in the teaching field. Workshops, seminars, literature review, and teacher conferences can provide the continuing education that these teachers need.

As stated earlier, continuing education can take the form of informal interaction with other teachers, particularly teachers who have more teaching experience. There are, however, other forms of continuing education equally valuable. Some of these are discussed briefly.

Mentoring

Teachers need to acquire classroom management skills that are not necessarily taught in formal education. For example, a teacher may need some guidance on how to handle a particular student situation. A more experienced teacher at the school may be able to share methods and approaches that he or she has developed over the course of time to help the new teacher. To the extent that such collaboration takes place, less experienced teachers may be able to minimize pitfalls that would make teaching more difficult.

Workshops and Conferences

In a workshop, the teacher steps into the role of student for a while. A leader may ask members at the workshop to undertake certain educational activities. Through the activities, teachers may learn new knowledge, sharpen various methodologies, and gain insight they did not have before. One advantage of attending a workshop is that teachers have the opportunity to strengthen their sense of community. Sharing ideas and learning from others not only makes teachers more knowledgeable in their subject areas, but it also reaffirms the belief that knowledge is by no means static but ever growing. Consequently, teachers can transmit to students their renewed enthusiasm about learning.

Books, Articles, and Periodicals

Reading is a source of continuing education. Using this vehicle, teachers may become informed about the views of others on education. One potential benefit of this type of continuing education is that the teacher may develop a deeper understanding of teaching methodologies and how other teachers structure educational experiences for students. In this book, you will find many discussions of algebra topics. These discussions provide educational opportunity in the content area of algebra as well as suggestions for sharing that knowledge with students.

Course Work

For teachers, learning has a multiplicative effect. For example, teachers who take courses at a community college can expand their own knowledge of a subject considerably. As an additional advantage, the more knowledge teachers acquire, the better prepared they are to teach students. Teachers who set aside time to continue their education are in fact investing in their own preparation as professionals and in the education of their students.

Summary

In this chapter, you have read a description of the task the middle grades mathematics teacher needs to accomplish. Our discussion has focused on important resources that can help teachers become better professionals. The purpose of our discussion is to invite teachers to reflect upon their roles as facilitators of learning experiences, role models, and members of a community of learners.

Teachers are invited to expand upon what has been said here and interpret it in the contexts of their individual teaching circumstances. Indeed, adapting what has been said here is another form of continuing education. As members of a learning community, teachers must develop or perhaps redefine their own identity as learners, information seekers, and problem solvers in the same manner as they expect their students to do. Understanding the task and how to accomplish it in particular settings can be considered a challenging problem-solving opportunity. The extent to which the educational enterprise succeeds is due in great part to the efforts of those who envision the task of teaching as a collaborative effort.

LINEAR EQUATIONS AND INEQUALITIES IN ONE VARIABLE

Making the Transition from Arithmetic to Algebra

In the study of whole numbers, students may encounter a problem like the one below.

> After 18 marbles are taken from a bag, there are 24 marbles still in it. How many marbles were in the bag before any marbles were taken out?

Students may also encounter a problem like this when they study what operation to choose to solve a problem or when they are studying addition of whole numbers with renaming.

Arithmetic Solution: Start with a bag that contains 24 marbles. Add back in 18 marbles. Therefore, add 18 to 24. The total number of marbles in the bag before any were taken was 42.

$$\begin{array}{r} {}^{1}1\,8 \\ +\ 2\,4 \\ \hline 4\,2 \end{array}$$

An alternative algebraic solution is shown below.

Algebraic Solution: Let m represent the number of marbles in the bag originally. Then the equation below represents the relationship between the beginning number of marbles, the number of marbles removed, and the final number of marbles in the bag.

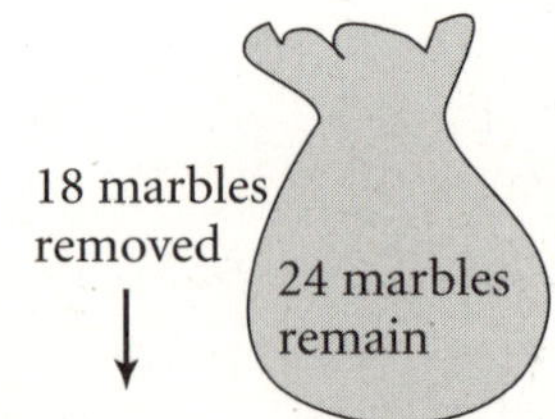

$$\text{beginning number} - \text{number removed} = \text{final number}$$
$$m \qquad - \qquad 18 \qquad = \qquad 24$$

Apply the Addition Property of Equality.

$$m - 18 + 18 = 24 + 18$$

By the Additive Inverse Property, drop $-18 + 18$ since that sum is 0.

$$m = 24 + 18$$
$$m = 42$$

The bag originally had 42 marbles in it.

Notice that after all the algebraic work is done, the student must still find the sum, $24 + 18$.

One might ask why it is necessary to learn how to solve equations using algebra if, when all is said, the final answer is found by using arithmetic. One part of the answer is that more complex problems cannot be solved by using arithmetic only. However, it is recommended that students learn how to solve simple problems by using algebraic methods before giving them more difficult problems to solve.

Open Sentences

When students evaluate an expression such as $\dfrac{3(4 + 5) + 5}{8}$, they learn to write an **equation.** Indeed, the first thing that a student writes after copying the expression is the sign for equality. To the right of the equal sign, they would write $\dfrac{3(9) + 5}{8}$. The student creates a true equation that involves numbers and operations on numbers. Notice that the equation does not involve any variable. The equation $\dfrac{3(4 + 5) + 5}{8} = \dfrac{3(9) + 5}{8}$ is always true.

equation: Two equivalent algebraic expressions separated by an equal sign.

In algebra, students encounter open sentences. An **open sentence** is one whose truth is not determined until more information is given. For example, the equation $m - 18 = 24$ is an open sentence. It is neither true nor false until a particular number replaces the variable m.

open sentence: A sentence whose truth is not determined until more information is given.

If m is replaced by 40, 50, and 60, a false equation results.

$40 - 18 = 22 \neq 24$ ✗ $50 - 18 = 32 \neq 24$ ✗ $60 - 18 = 42 \neq 24$ ✗

There is only one replacement for m that will make the equation true. That value of m is 42.

$$42 - 18 = 24 \checkmark$$

One of the primary objectives in an introductory algebra course is developing methods to find the numbers that make an open sentence true. These numbers are called the *solutions* to the open sentence.

One purpose of checking a solution, once one is found, is to verify that the number found makes the given open sentence true. Numbers that make a statement false do play a role in deciding whether a given statement is true for all replacement values of a variable.

Consider the statement below. Is it true for all natural numbers?

Every positive multiple of 3 is an odd number.

Let *n* represent any natural number. If $n = 1$, then $3(1) = 3$, an odd number. If $n = 2$, then $3(2) = 6$, an even number. The statement is true when $n = 1$ but false when $n = 2$. Replacement of the variable by a number that makes the statement false helps you decide whether a statement is true for all natural numbers. The given statement is not always true.

Assessing the truth of a statement involving a variable may not help solve a real-world problem, but it does help students practice their logical reasoning.

Solving Simple Equations

The illustration at right suggests that if weights *a* and *b* cause the scale to balance, then adding the same weight, *c*, onto each pan of the scale will cause the scale to continue to balance. Indeed, this is true. You can formalize this observation as the *Addition Property of Equality*.

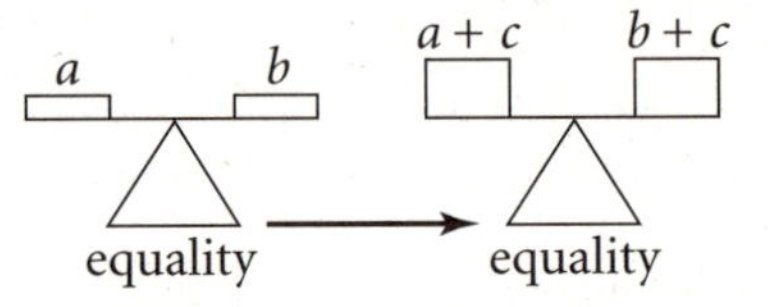

If *a*, *b*, and *c* are real numbers and $a = b$, then $a + c = b + c$.

The Addition Property of Equality was the key element in the solution of the problem about marbles in a bag on page 18.

Consider the problem below.

After 18 marbles are put into a bag, there are 42 marbles in the bag. How many marbles were in the bag before any marbles were added?

This problem cannot be solved by addition. Consider the explanation below.

Solution: Let *m* represent the original number of marbles in the bag. Then the equation below represents the relationship between the original number of marbles, the number of marbles added, and the final number of marbles.

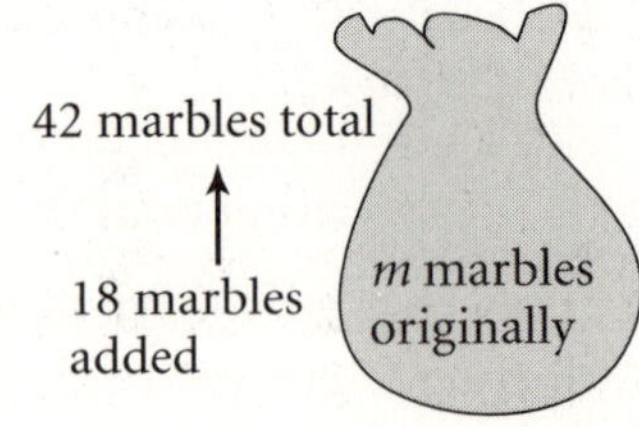

$$m + 18 = 42$$

Subtract 18 from each side of the equation.

$$m + 18 - 18 = 42 - 18$$

By the Additive Inverse Property, drop $18 - 18$ since it equals 0.

$$m = 42 - 18$$
$$m = 24$$

The bag originally had 24 marbles in it.

You can use addition to check the solution found. Start with 24 marbles and add 18 marbles. You will have 42 marbles in all.

This problem reveals another method that can be used to solve a simple equation. This method requires the application of the *Subtraction Property of Equality* illustrated at right.

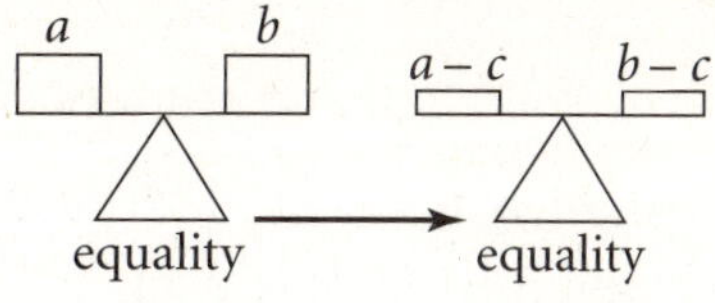

If a, b, and c are real numbers and $a = b$, then $a - c = b - c$.

Careful reading will help you decide whether an equation requires addition or subtraction in order to solve it.

The following two problems look similar to each other but are solved differently. Neither addition nor subtraction is used as a method of solution.

  I How many chips should a teacher set aside so that each of 20 children will have 18 chips?

  II How many chips should a teacher give each of 20 students if the teacher has 360 chips to distribute and each student is to receive an equal number of chips?

If you write an equation to solve each problem, you might reason as follows.

Problem I: Let s represent the number of chips the teacher needs to set aside. In the diagram at right, you can see a representation of the distribution. Notice that you can expect s to be larger than 18.

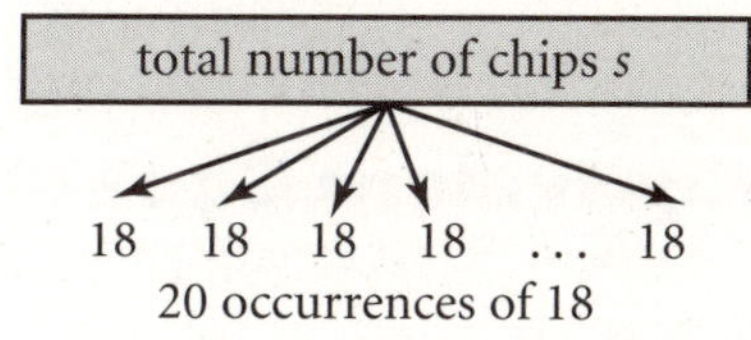

$$\frac{\text{total number of chips}}{\text{number of students}} = \text{number of chips per student}$$

$$\frac{s}{20} = 18 \quad \leftarrow \quad \textit{Multiply 18 by 20.}$$

$$s = 20 \times 18 = 360$$

The teacher needs to set aside 360 chips.

Problem II: Let t represent the number of chips each student would receive.

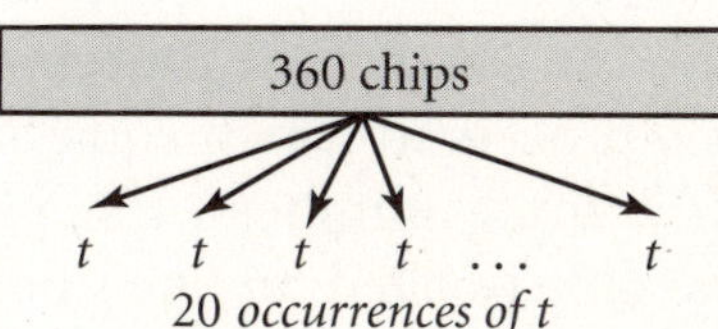

number of students
    $\times$ number of chips per student
       $=$ total number of chips

$$20 \times t = 360 \quad \leftarrow \quad \textit{Divide 360 by 20.}$$

$$t = \frac{360}{20} = 18$$

Each student will receive 18 chips.

To solve Problem I, multiply each side of the equation by 20. To solve Problem II, divide each side of the equation by 20.

These problems suggest two more properties of equality that can be used to solve simple equations. The first property below is used to solve Problem I. The second property below is used to solve Problem II.

Multiplication Property of Equality
If a, b, and c are real numbers and $a = b$, then $ac = bc$.

Division Property of Equality
If a, b, and c are real numbers, $c \neq 0$, and $a = b$, then $\dfrac{a}{c} = \dfrac{b}{c}$

From a study of the four problems discussed previously, you can say that addition and subtraction are inverse operations and that multiplication and division are inverse operations. These statements are equivalent to saying that a number added to its opposite is 0 and that a number times its reciprocal is 1.

$$-18 + 18 = 0 \qquad 18 - 18 = 0 \qquad 20 \times \frac{1}{20} = 1 \qquad \frac{1}{20} \times 20 = 1$$

Equivalent Equation

When you apply a property of equality to solve an equation, you transform a given equation into an *equivalent equation,* that is, an equation having the same solution(s) as the given equation.

$$
\begin{array}{lll}
\text{given equation} \rightarrow & \text{equivalent equation} & \rightarrow \text{equivalent equation} \\
m - 18 = 24 \ \rightarrow & m - 18 + 18 = 24 + 18 \ \rightarrow & m = 42 \\
m + 18 = 42 \ \rightarrow & m + 18 - 18 = 42 - 18 \ \rightarrow & m = 24 \\
\dfrac{s}{20} = 18 \ \rightarrow & 20 \times \dfrac{s}{20} = 20 \times 18 \ \rightarrow & s = 360 \\
20t = 360 \ \rightarrow & 20t \div 20 = 360 \div 20 \ \rightarrow & t = 18
\end{array}
$$

You can say that the process of solving an equation containing one variable is a process in which equations are transformed into equivalent equations in a finite number of steps. The resulting equivalent equation has the variable isolated on one side of the equation (usually the left side) and one number on the other side of the equation (usually the right side).

final equivalent equation: variable name = number

The process of solving an equation requires making decisions. For example, to apply a property of equality productively, the student must choose what number to add, subtract, multiply by, or divide by. The examples below illustrate that one can get an equivalent equation without finding a solution.

$$m - 18 = 24$$

Add 10.	Add 20.	Add 18.
$m - 18 = 24$	$m - 18 = 24$	$m - 18 = 24$
$m - 18 + 10 = 24 + 10$	$m - 18 + 20 = 24 + 20$	$m - 18 + 18 = 24 + 18$
$m - 8 = 34$	$m + 2 = 44$	$m = 42$
solution not found	solution not found	solution found ✔

Notice in the reasoning above that all choices for a number to add are correct and the Addition Property of Equality is applied correctly. However, only the third choice, Add 18, is *both* correct and productive.

Each of the equations below are equivalent to one another.

$$x + 3 = 7 \qquad 3 + x = 7 \qquad 7 = x + 3$$

Since addition is commutative: $\quad 3 + x = 7 \quad \Leftrightarrow \quad x + 3 = 7$
Since equality is symmetric: $\quad 7 = x + 3 \quad \Leftrightarrow \quad x + 3 = 7$

Therefore, regardless of which of these three equations are given to you, you should be able to solve all of them by subtracting 3 from x and subtracting 3 from 7.

In textbooks, students are given equations in which the variable is not on the left side of the equation (Examples **3** and **4**) and equations in which a number appears at the left of one side of the variable (Examples **2** and **3**). The idea is to make students see that, although equations may look different, they are in fact the same from the point of view of solving equations. Students are also given equations in which the variable name changes from one equation to another. All the equations below are solved in exactly the same way.

1 $a + 12 = 30$ $\qquad$ **2** $12 + d = 30$ $\qquad$ **3** $30 = 18 + w$ $\qquad$ **4** $30 = y + 18$

Students often ask for help when it comes to choosing the appropriate operation to do. Teachers can help students make the correct choice by asking students the following questions.

- Is a number added to the variable? If the answer is Yes, take that number away from each side of the equation.

- Is a number subtracted from the variable? If the answer is Yes, add that number to each side of the equation.

- Is a number multiplying the variable? If the answer is Yes, divide both sides of the equation by that number.

- Is a number dividing the variable? If the answer is Yes, multiply both sides of the equation by that number.

Solving Involved Equations

After students learn how to solve a simple equation in one step, they learn how to chain various properties together to make equivalent equations that eventually lead to the solution of a complicated equation. Of particular concern is helping students deal with equations that require multiple steps to find a solution.

Consider $3x + 5 = 17$.

$3x + 5 = 17$	*given equation*
$3x + 5 - 5 = 17 - 5$	*Subtraction Property of Equality*
$3x + 0 = 12$	*Additive Inverse Property*
$3x = 12$	*Property of Zero*
$x = 4$	*Division Property of Equality*

When students are taught how to solve $3x + 5 = 17$, they are taught to subtract 5 from each side of the equation, then divide each side of the resulting equivalent equation by 3. Thus, students think of $3x + 5 = 17$ as a *two-step equation*.

$$\text{Step 1} \qquad \text{Step 2}$$
$$3x + 5 = 17 \quad \rightarrow \quad 3x = 12 \quad \rightarrow \quad x = 4$$

An equation may require multiple steps. For example:

Consider $3m + 8 - 5m = 9 + 4m + 29$.

Simplify the left and right sides of the equation separately.

$$3m + 8 - 5m = \mathbf{-2m + 8} \qquad 9 + 4m + 29 = \mathbf{4m + 38}$$

Now solve an equivalent equation.

$-2m + 8 = 4m + 38$	
$-2m = 4m + 30$	*Subtract 8 from each side.*
$-6m = 30$	*Subtract 4m from each side.*
$m = -5$	*Divide each side by -6.*

Therefore, the solution to $3m + 8 - 5m = 9 + 4m + 29$ is -5.

A schematic view of the solution process for solving $3m + 8 - 5m = 9 + 4m + 29$ is shown below. Notice how algebraic methods are chained together in a finite number of organized steps to arrive at the solution.

Simplify each side of the equation.

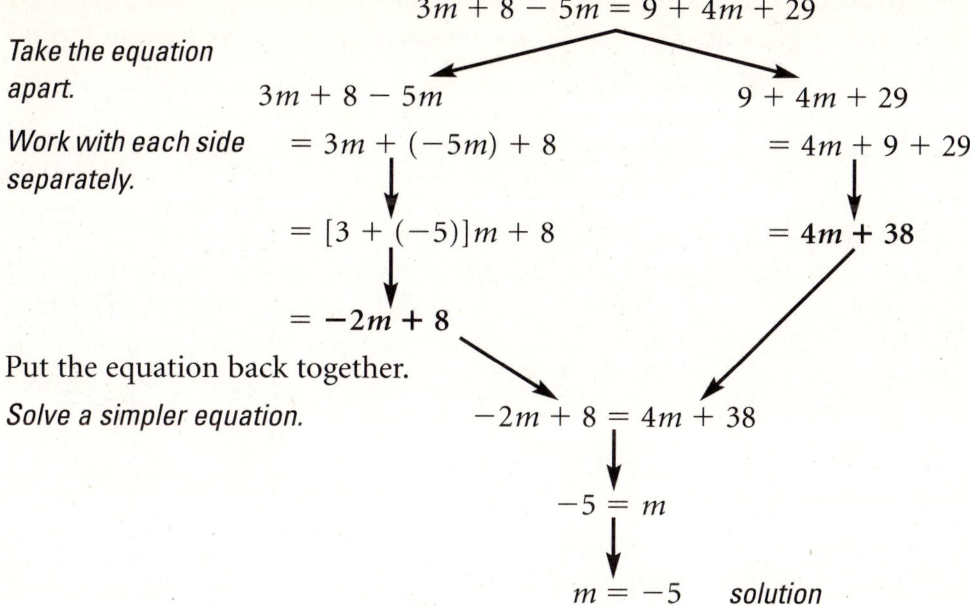

Some textbook exercises display a check on the solution found. A check on the work and the final result reminds students that the intermediate equations used to arrive at a solution are equivalent. Students should be encouraged to check their solutions before turning in their work. Checking a solution also helps the student develop a cautious and careful attitude to their work. Finally, a check can help the teacher diagnose where a student went wrong in the event that the solution found is not correct.

Consider the following solution to $3m + 8 - 5m = 9 + 4m + 29$.

$$3m + 8 - 5m = 9 + 4m + 29$$
$$3m - 5m + 8 = 4m + 9 + 29$$
$$-2m + 8 = 4m + 38$$
$$-2m = 4m + 30$$
$$2m = 30 \quad \textbf{✗}$$
$$m = 15 \quad \textbf{✗}$$

Check: $3(15) + 8 - 5(15) = -22 \quad 9 + 4(15) + 29 = 98 \quad -22 = 98 \; \textbf{✗}$

The check indicates that 15 is not the solution to the given equation. Something went wrong. To find where an error was made, work backward. Check 15 in each equation preceding $m = 15$ working from $2m = 30$ up the chain of equations.

$$m = 15$$
$$2(15) = 30 \quad \textbf{✔}$$
$$-2(15) = 4(15) + 30 \quad \textbf{✗}$$

A mistake was made in going from $-2m = 4m + 30$ to the next equation. The student added $4m$ to the left side of the equation to get $2m$ but added $-4m$ to the right side of it. Rather than writing $2m = 30$, the student should have written $-6m = 30$.

Developing equation-solving skills to go from solving an equation like $m - 18 = 24$ to solving an equation like $3x + 5 = 17$ to an equation as complicated as $3m + 8 - 5m = 9 + 4m + 29$ may take several weeks of mathematics education and practice.

The final step in the student's education about solving linear equations in one variable is to help students recognize that:

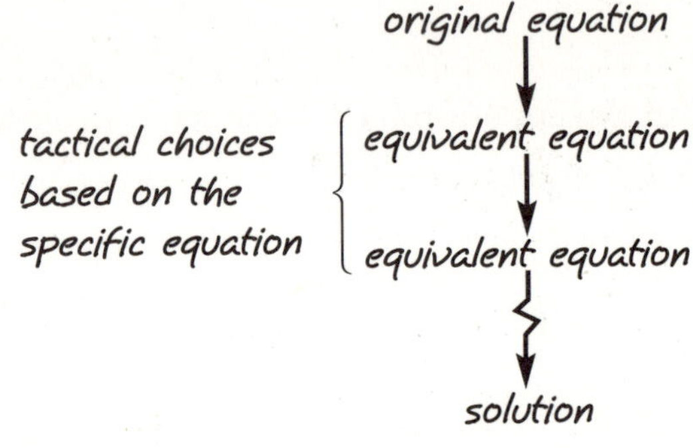

- the overall solution strategy is the same, namely, transform one equation into a series of equivalent equations that eventually give the solution, and

- each equation is specific in that the tactics for solving one equation may be the same as or different from those used to solve another equation.

Solving a Linear Equation in One Variable

Every linear equation in one variable can be solved by use of the properties of real numbers together with the properties of equality in some number of steps.

When students gain experience in solving linear equations in one variable, they will be prepared for any of the solution possibilities. For the most part, students are given equations that have exactly one solution.

Counting Solutions to Linear Equations in One Variable

Every linear equation in one variable and involving real numbers has one of the following solution possibilities.

no solution $\Leftrightarrow$ empty number line
exactly one solution $\Leftrightarrow$ one point on the number line
all real numbers as solutions $\Leftrightarrow$ the entire number line

The solution possibility will show itself in the solution process.

Consider $2(x + 4) = 2x + 7$.

$$2x + 8 = 2x + 7$$
$$8 = 7 \qquad \textit{no solution}$$

Consider $2(x + 4) = 3x + 7$.

$$2x + 8 = 3x + 7$$
$$x = 1 \qquad \textit{exactly one solution}$$

Consider $2(x + 2) + 4 = 2x + 8$.

$$2x + 4 + 4 = 2x + 8$$
$$2x + 8 = 2x + 8$$
$$0 = 0 \qquad \textit{all real numbers}$$

Introducing Linear Inequalities in One Variable

What is an inequality? Simply put, an **inequality** is a statement that two quantities are not equal.

equation   inequality

$$4 = \frac{20}{5} \qquad\qquad 4 \neq \frac{21}{5}$$

If two real numbers are not equal, then one of them is less than ($<$) the other and one of them is greater than ($>$) another. When you compare 4 with $\frac{21}{5}$, or $4\frac{1}{5}$, you can say that $4 < \frac{21}{5}$ and you can also say that $\frac{21}{5} > 4$.

The skill of comparing two numbers involves deciding which one is greater and which one is smaller. The skill of ordering a collection of numbers involves placing them in order from least to greatest.

A **linear inequality** in one variable is any inequality that involves a linear expression in that variable. Some inequalities include the equal sign (Examples **2** and **4**). These are called *weak inequalities*. Inequalities that do not contain the equal sign are called *strict inequalities*. The following are examples of inequalities involving one variable.

  1 $x + 5 > 7$ **2** $2n - 6 \geq -3$ **3** $4 + 5z < 3z$ **4** $3(d - 2) \leq 5$

A *solution* to a linear inequality in one variable is any number that makes the inequality true. The definition is similar to that of a solution to a linear equation. There is, however, a significant difference.

Consider $x + 5 > 7$.

   If $x = 0$, then $5 > 7$. This inequality is false. So, 0 is not a solution.

inequality: A statement that two expressions are not equal. Inequalities contain one of the following signs: $<, >, \leq, \geq,$ or $\neq$.

linear inequality: An inequality whose form is like that of a linear equation with the equal sign replaced by an inequality symbol.

If $x = 3$, then $8 > 7$. This inequality is true. So, 8 is a solution.

If $x = 4$, then $9 > 7$. This inequality is true. So, 9 is also a solution.

If $x = 5$, then $10 > 7$. This inequality is true. So, 10 is also a solution.

What becomes clear from this example is that $x + 5 > 7$ has at least three distinct solutions and that not all numbers are solutions.

How can a linear inequality in one variable be solved? The properties of addition and multiplication can still be used to help solve an inequality. However, the properties of equality are no longer applicable. To help answer the question, consider the problem below.

A motorist is driving along a highway at 45 miles per hour. Speed limit on the highway is 60 miles per hour. By how much may the motorist increase speed and still drive at or under the speed limit?

You can reason as follows.

$$\text{current speed} + \text{increase in speed} \le \text{speed limit}$$

$$45 + \text{increase in speed} \le 60$$

Using experimentation, you can find that any increase in speed from no increase at all to as much as 15 miles per hour will allow the motorist to stay at or under the speed limit. (Notice that 15 is exactly what you get when you subtract 45 from 60.)

The solution above suggests that you can generalize the work in the following statement.

If I subtract the same quantity from each side of an inequality, I will get another inequality that is as true as the given inequality.

This statement is similar to the Subtraction Property of Equality but it applies to inequalities. Its name is the *Subtraction Property of Inequality*. In all, there are four properties of inequality that will be useful in solving inequalities.

Properties of Inequality

Subtracting the same quantity from each side of a true inequality will give another true inequality. It is worth spending some time reflecting on this. The diagram on the next page shows two collections of pencils. Collection A has 3 pencils more than Collection B. The diagram also shows 4 pencils removed from each collection. Notice that after the removal, Collection A still has 3 pencils more than Collection B.

Collection A: 9 pencils   Collection B: 6 pencils

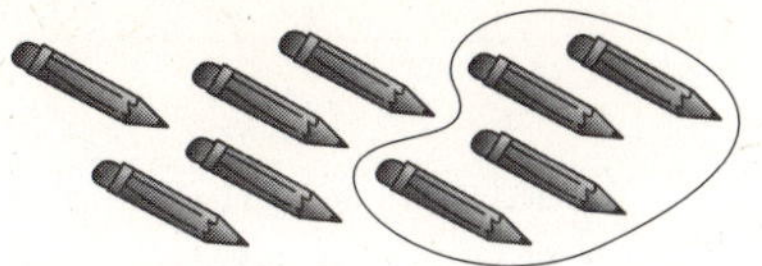   

Remove 4 pencils.   Remove 4 pencils.

It is not possible to represent the removal of 12 pencils from each collection since neither collection has that many pencils. However, you can generalize to say that if a and b represent numbers and $a < b$, then, whatever c is, you can truthfully state that $a - c < b - c$. This is the Subtraction Property of Inequality.

Using models like the one above, you can also write a Multiplication Property of Inequality as long as the multiplier is a positive number. What can one conclude if the multiplier is a negative number?

Consider $2 < 3$ and the products $2(-3)$ and $3(-3)$.

The number line below shows $2 < 3$ and the products $2(-3)$ and $3(-3)$.

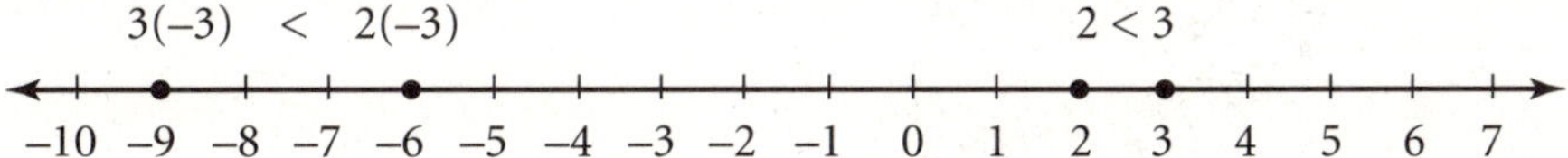

The inequality relationship is reversed. This happens when a negative multiplier is used in an inequality.

In all, there are four properties of inequality.

Properties of Inequality

Let a, b, and c represent real numbers.

Addition:	If $a < b$, then $a + c < b + c$.
Subtraction:	If $a < b$, then $a - c < b - c$.
Multiplication:	If $a < b$ and $c > 0$, then $ac < bc$.
	If $a < b$ and $c < 0$, then $ac > bc$.
Division:	If $a < b$ and $c > 0$, then $\dfrac{a}{c} < \dfrac{b}{c}$.
	If $a < b$ and $c < 0$, then $\dfrac{a}{c} > \dfrac{b}{c}$.

Similar statements apply if $<$ is replaced by $>$, $\le$, or $\ge$.

Solving Linear Inequalities in One Variable

Students follow the same procedure as with equations when they learn how to solve linear inequalities in one variable. At first, students learn to apply one property of inequality to find a solution.

Consider $3u \geq 6$.

$$3u \geq 6$$

$$\frac{3u}{3} \geq \frac{6}{3} \qquad \textit{Divide each side of the inequality by 3.}$$

$$u \geq 2$$

Gradually, students learn to apply two properties of inequality to solve a linear inequality in one variable.

Consider $4 + 5z < 3z$.

$$4 + 5z < 3z$$

$$4 < -2z \qquad \textit{Subtract 5z from each side.}$$
$$\textit{Divide each side by} -2 \textit{ and reverse the inequality}$$
$$\textit{symbol.}$$

$$z < -2$$

Students then progress, as they did in the study of linear equations in one variable, to the point where they use multiple properties of inequality to arrive at a solution. You can emphasize the similar procedures involved in solving linear equations in one variable and linear inequalities in one variable. However, make sure that students remember to reverse an inequality symbol when they multiply or divide by a negative number.

Consider $3m + 8 - 5m \geq 9 + 4m + 29$.

Simplify each side of the inequality.

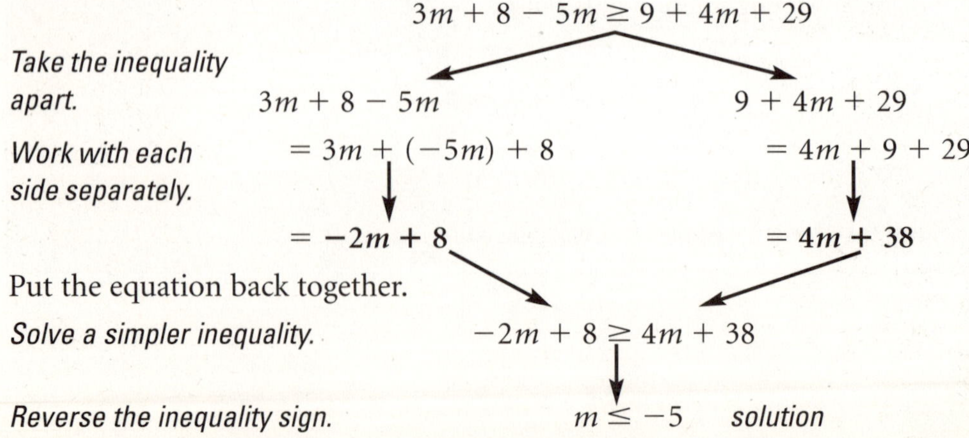

Put the equation back together.

Reverse the inequality sign.

When students are given an equation in one variable to solve, the solution is usually one number. Students are not asked to represent the solution on a number line. However, students are often asked to represent a solution to an inequality in one variable on a number line because that solution includes more than one point.

For example, the solution to $3u \geq 6$ is represented by an arrow, or ray, starting at 2 and pointing to the right. The closed circle at the endpoint of the ray indicates that 2 is in the solution.

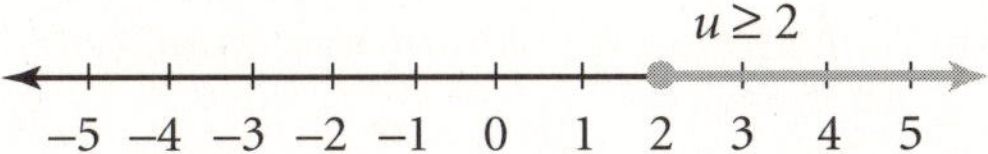

The solution to $4 + 5z < 3z$ is represented by an arrow, or ray, starting at -2 and pointing to the left. The open circle at the endpoint of the ray indicates that -2 is not in the solution.

Solving a Linear Inequality in One Variable

Every linear inequality in one variable can be solved by the application of the properties of real numbers and the properties of inequality. Most often, the solution is represented by a ray on the number line. The endpoint of the ray is either closed ($\leq$, $\geq$) or open ($<$, $>$).

After students gain experience in solving linear inequalities in one variable, they will be better prepared to recognize any of the solution possibilities.

Counting Solutions to Linear Equations in One Variable

Every linear inequality in one variable and involving real numbers has one of the following solution possibilities.

no solution

an open or closed ray pointing left or right

all real numbers as solutions

The solution possibility will show itself in the solution process.

Consider $2(x + 3) > 2x + 7$.

$6 > 7$ *no solution*

Consider $2(x + 4) \leq 3x + 7$.

$x \geq 1$ *a ray with its endpoint closed*

Consider $2(x + 2) + 4 \geq 2x + 8$.

$0 \geq 0$ *all real numbers*

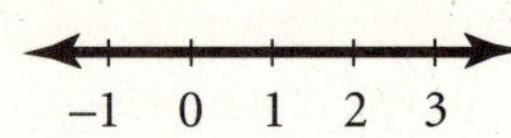

If you select any number, r, on the number line, you divide the number line into three components, the point representing r, the points corresponding to numbers less than r $(x < r)$, and points corresponding to numbers greater than r $(x > r)$.

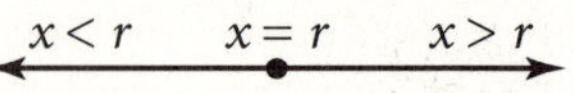

This insight provides the following strategy for solving a linear inequality in one variable by using a linear equation in one variable. An example will illustrate this strategy.

Consider $4 + 5z < 3z$.

$$4 + 5z < 3z$$

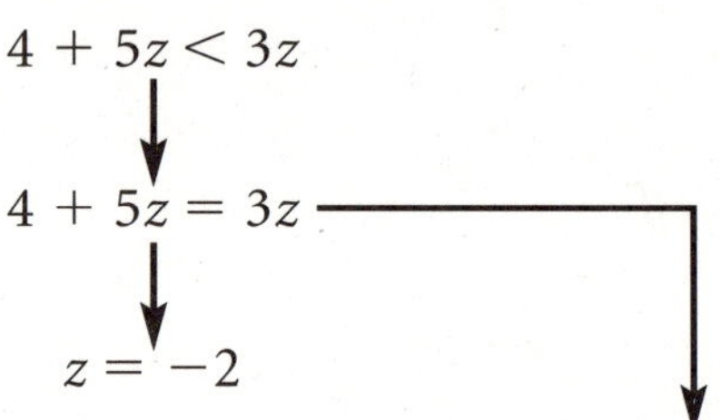

Replace the inequality sign with an equal sign.

$$4 + 5z = 3z$$

Solve the equation.

$$z = -2$$

Locate -2 *on a number line. Draw an open circle.*

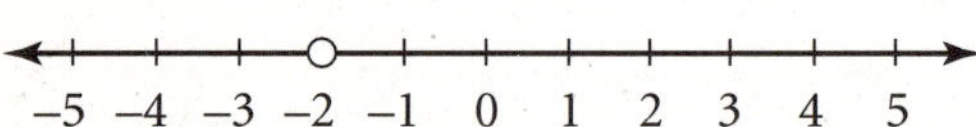

Test a number such as 0 to see if it is a solution.

$$4 + 5(0) = 4 \qquad 3(0) = 0$$
$$4 \neq 0$$

Therefore, 0 is not a solution.

Draw the ray starting at -2 *but in the direction opposite to 0.*

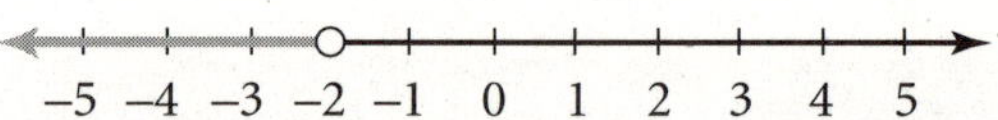

The procedure that outlines the solution to $4 + 5z < 3z$ points out the following important facts.

- Equation-solving skills are not restricted to the study of equations. Indeed, when a student solves an inequality by using an equation, the student transfers his or her equation-solving abilities across related but separate domains. Transfer is a measure of understanding.

- A number either is or is not a solution to a given inequality. Using a number is a logical test to find which ray, a left-pointed ray or a right-pointed ray, is the solution.

- Using an equation to help solve an inequality is a powerful problem-solving strategy. After all, the solution to the equation gives the boundary line for the solution to the inequality. Consequently, this strategy helps students become better problem solvers.

One of the reasons that students need to study linear inequalities in one variable is that they are used in many real-world problems.

Consider this problem.

> A small-business owner purchased a computer system recently for $12,500. Each year, the system will lose $450 in value as it depreciates. During what years, starting at the year 2000, will the value of the computer system be at least $5750?

> To solve this problem, you will need to devise a plan.

❶ Let t represent the number of years. Then $450t$ will represent the loss in value over t years and $12,500 - 450t$ will represent the value of the computer system after t years.

Write a relationship involving known and unknown quantities. To decide which inequality sign to use, make sure that students understand that the phrase "at least $5750" means "equal or greater than $5750."

$$\text{values after } t \text{ years is } \$5750 \text{ or more}$$
$$12,500 - 450t \geq 5750$$

❷ Solve the inequality.

$$12,500 - 450t \geq 5750$$
$$-450t \geq -6750$$
$$t \leq 15$$

❸ Answer the question.

The value of the computer system will be at least $5750 in the years from 2000 to 2015 inclusive.

It is worth reflecting on the solution above.

- Notice, for example, that the solution $t \leq 15$ includes all real numbers less than or equal to 15. The solution contains negative numbers. Obviously, if t is a negative number, then the year would be a year before 2000. The implicit restriction $t \geq 0$ adds meaning and logic to the context of this problem.

- Notice also that the solution to the inequality is not the answer to the question. Instead the problem solver needs to interpret the solution to see how it helps answer the question.

Solving Compound Linear Inequalities

The problem involving the depreciation of a computer system gives two related inequalities. The inequality $t \leq 15$ comes from the solution and $t \geq 0$ is an implicit restriction.

$$t \geq 0 \text{ and } t \leq 15$$

A timeline that represents the answer to the computer system problem is shown below.

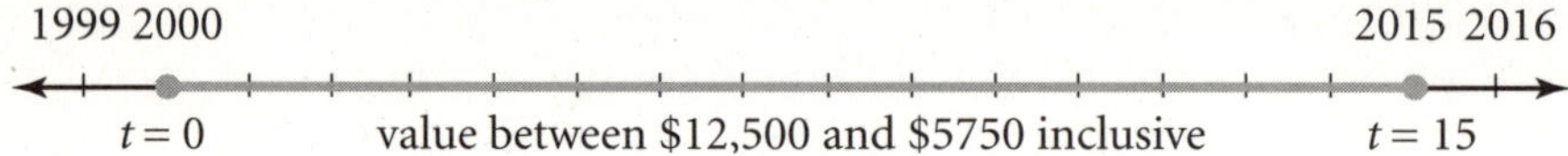

Notice that the representation of the answer to the question is a *line segment,* a part of the number line with two distinct endpoints.

compound inequality: Two inequalities that are combined into one statement by the word *and* or *or.*

A **compound inequality** is a pair of linear inequalities in one variable linked by the word *and* or by the word *or.* Solving a compound inequality will take us back to methods for solving a single inequality in one variable.

Consider $x \geq -2$ and $x < 4$.

To represent the solution to the pair of inequalities, make a number line for each inequality. Place one number line above the other. Then make a third number line showing the line segment that is common to both lines.

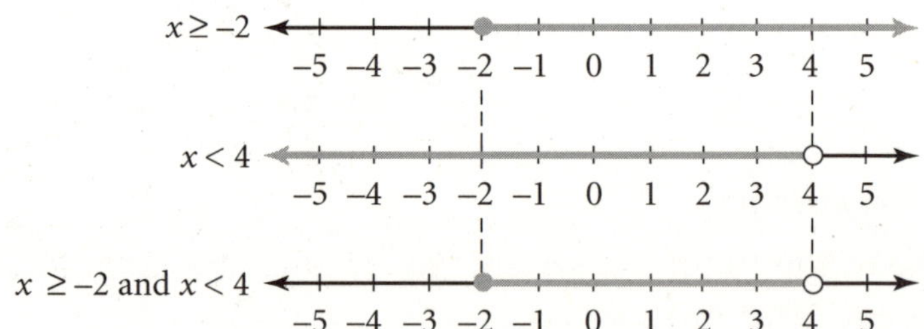

Consider $x < -2$ or $x \geq 4$.

Make a number line for each inequality. Then make a third number line showing the line segments from either of the two number lines.

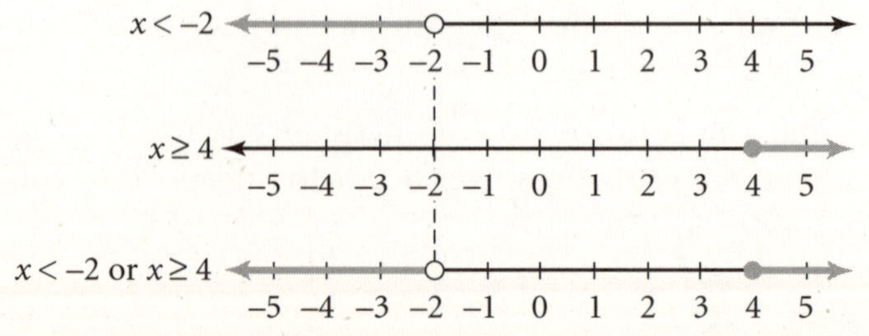

You can motivate your students to achieve more by showing them how far they have come in their education as mathematical thinkers. It is important for students to realize that their accumulated knowledge of skills has made them better problem solvers and more efficient thinkers. The schematic representation below shows how a large collection of skills come into play in the solution to a compound inequality.

Consider $2x - 1 < 5$ and $2x + 9 \geq 7$.

Take the compound inequality apart into two inequalities.

Work on each one separately.

Bring it all back together.

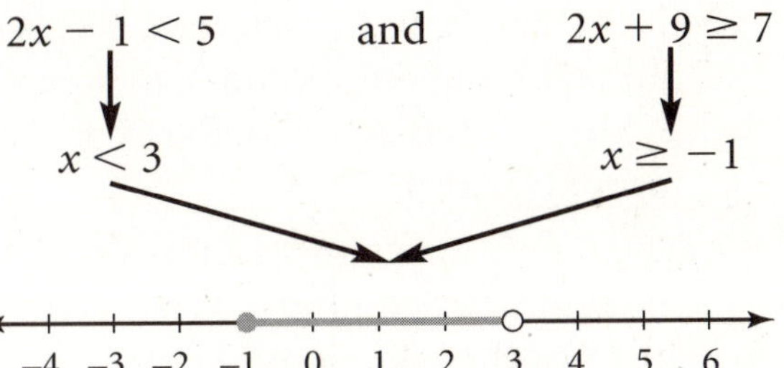

Encourage students to participate at the whiteboard, explain each step taken, and graph the solution to an inequality. Guided participation allows students to practice their procedural knowledge and enhance their understanding.

Absolute Value

Recall that the **absolute value** of a real number x, denoted $|x|$, is defined as the distance between the graph of that number and 0 on the number line. In other words, the absolute value is a *distance function* whose domain is the set of all real numbers and whose range is the set of nonnegative real numbers.

> **absolute value:** The absolute value of a real number x is the distance from x to 0 on a number line. The symbol $|x|$ means the absolute value of x.

In the diagram below, notice that 3 and -3 are both 3 units from 0.

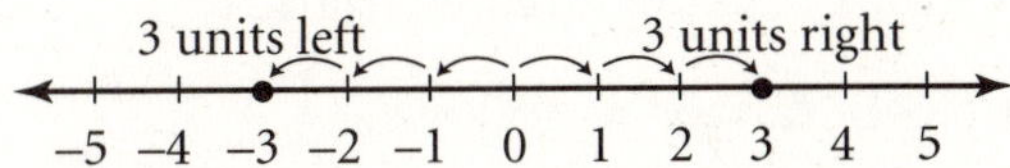

Both 3 and -3 have absolute value 3.

$$|-3| = 3 \text{ and } |3| = 3$$

One application of absolute value is in the addition of integers. For example, the sum $-7 + 4$ is found by computing $|-7| - |4|$ and then attaching $(-)$ to the result. So, $-7 + 4 = -3$. In algebra, absolute value is also connected to solving equations and inequalities. Beginning algebra books usually contain at least one lesson involving absolute-value equations and inequalities.

Consider again the number line on page 35. The points at -3 and 3 divide the number line into sections.

- The points at -3 and 3 are the only ones that are exactly 3 units from 0. If x is either of these numbers, you can say that $|x| = 3$.

- One of the sections on the number line is the set of points between -3 and 3. What can be said of these points? Clearly, if you choose a point between -3 and 3, you will have chosen a point whose distance from 0 is less than 3. For example, 1 is closer to 0 than is -3 or 3. If x is any number between -3 and 3, then its distance from 0 is less than 3. That is, $x > -3$ and simultaneously $x < 3$. You can express this in compact form as $|x| < 3$.

- The points in the two sections to left of -3 and to the right of 3 have something in common. What can be said about them? Any point in either of these sections is farther from 0 than is -3 or 3. For example, -10 and 12 are farther from 0 than are -3 and 3. If x is any number less than -3 or any number greater than 3, then $x < -3$ or $x > 3$. You can express this in compact form as $|x| > 3$.

As any other concept, the concept of absolute value requires analytical thinking. To make general statements, students must engage in analytical thinking. Suppose that -3 and 3 are replaced by $-a$ and a, where a is positive. What follows is the result of analytical thinking.

- The equation $|x| = a$ tells you what numbers are exactly a units from 0 on the number line. If $a = 3$, then you may conclude that x must equal 3 or x must equal -3.

- The inequality $|x| < a$ tells you what numbers are less than a units from 0 on the number line. If $a = 3$, then you may conclude that x must be between -3 and 3.

- The inequality $|x| > a$ tells you what numbers are more than a units from 0 on the number line. If $a = 3$, then you may conclude that x must be less than -3 or x must be greater than 3.

These results can be generalized as follows.

$|x| = a$, $a > 0$; solution: a pair of points each a units from 0

$|x| < a$, $a > 0$; solution: a line segment whose endpoints are at a and $-a$, not including a and $-a$

$|x| > a$, $a > 0$; solution: the ray starting at $-a$ and pointing left along with the ray starting at a and pointing right

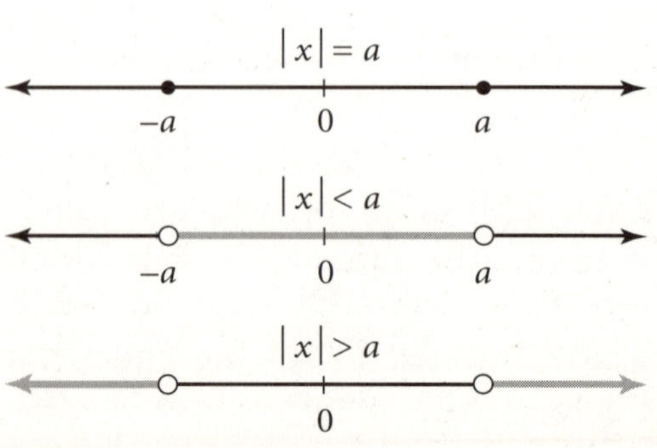

How can this analysis help solve an equation involving absolute value? You can argue that if $|x| = 5$, then $x = -5$ or else $x = 5$. You can solve more complicated equations such as $|2y - 1| = 5$ by using what you already know.

Consider $|2y - 1| = 5$.

The number y causes the more complicated number $2y - 1$ to be exactly 5 units from 0 on the number line. This is what absolute value means. The only numbers that are 5 units from 0 are -5 and 5. Therefore, you are looking for a number, y, that makes $2y - 1$ equal to -5 or that makes $2y - 1$ equal to 5.

$$y \text{ makes } 2y - 1 \text{ equal to } -5 \quad \text{or} \quad y \text{ makes } 2y - 1 \text{ equal to } 5$$
$$2y - 1 = -5 \qquad \text{or} \qquad 2y - 1 = 5$$

Notice that the two equations above are both linear equations in one variable. The next step is to solve each equation for y.

$$2y - 1 = -5 \qquad \text{or} \qquad 2y - 1 = 5$$
$$2y = -4 \qquad \text{or} \qquad 2y = 6$$
$$y = -2 \qquad \text{or} \qquad y = 3$$

The solution to the problem is now complete.

You can reflect on the steps of the solution to the problem above and devise a method for solving **absolute-value equations**. This method is outlined below.

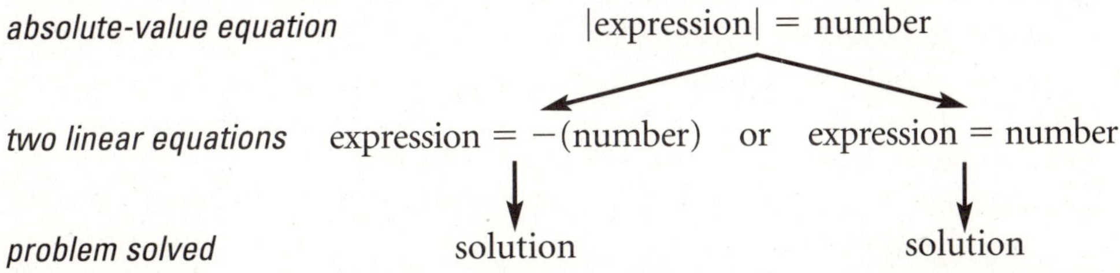

absolute-value equation: An equation that includes an absolute value. Absolute-value equations have either two solutions or none.

The strategy of transforming an equation into one or more equations is a strategy used very often in mathematics. It is used here and will be used in later contexts when more unfamiliar problems are encountered.

What about solving an inequality involving absolute value? Can the method above be modified to solve an inequality involving ($<$) or ($\le$)? The answer is Yes. The modifications are not complicated and the modified model is shown below.

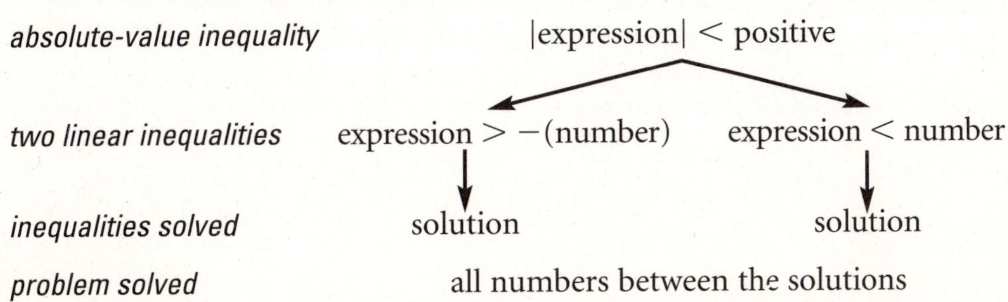

Now think about how well this method works when it comes to solving an inequality like $|2(x - 5)| < 2$.

Consider $|2(x - 5)| < 2$.

absolute-value inequality	$|2(x - 5)| < 2$
two linear inequalities	$2(x - 5) > -2$ and $2(x - 5) < 2$
	$2x - 10 > -2$ and $2x - 10 < 2$
	$2x > 8$ and $2x < 12$
inequalities solved	$x > 4$ and $x < 6$
problem solved	all numbers between 4 and 6

As you can see, the strategy works perfectly well. How can you help students develop problem-solving strategies like those explained here? The textbook by itself is probably not sufficient to elaborate on all the thinking shown here.

Shown below are some options that teachers might use.

- Create a handout that explains in detail how to solve an equation like $|2y - 1| = 5$. Have students place a sheet of paper over the handout and uncover one step at a time as you guide them through the solution. This approach has the advantage that students do not see the entire solution at one glance.

- Initiate a whiteboard discussion that unfolds one step at a time as students brainstorm over what step should come next. That step is written on the whiteboard once students agree that it is a sensible next step. During the discussion, students discover for themselves that equation-solving skills return and become useful in solving a new problem. Students will need some guiding questions to help them understand the logic in each step.

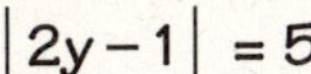

$$|2y - 1| = 5$$

Class: What does absolute value tell you about the distance between $2y - 1$ and O?
Response: That distance must be 5 units.

Class: Where would you place $2y - 1$ on the number line?
Response: Place $2y - 1$ at -5 and at 5.

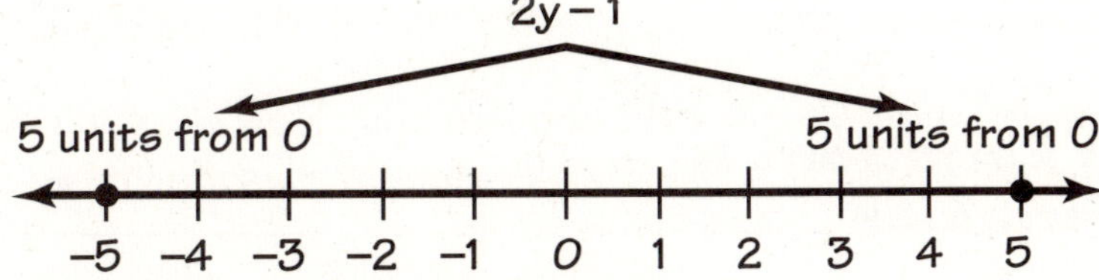

Class: Where do we go from here?
Response: Solve two equations.

$$2y - 1 = -5 \qquad\qquad 2y - 1 = 5$$

Class: How do we solve these equations?
Response: Use properties of equality.

$$y = -2 \qquad\qquad y = 3$$

However the exploration of absolute value unfolds, keep in mind that the major objective is to help students realize that previously learned skills are used to solve new problems.

The question at right can be used to remind students to use what they already know about problem solving when they encounter a new or unfamiliar problem.

new problem
How can I use what I learned before together with my thinking skills to solve a new problem?

LINEAR EQUATIONS AND INEQUALITIES IN TWO VARIABLES

Linear Equations in Two Variables

The equation $3x = 6$ is an equation in one **variable,** x, that has exactly one solution, 2. What would happen if a second variable, y, was introduced?

$$3x + y = 6$$

The equation above is a **linear equation in two variables.** The study of two-variable equations is important because many real-world problems involve two variables instead of just one. A solution to a two-variable equation cannot be a single number.

Consider the equation $3x + y = 6$.

$$\text{If } x = -1, \quad \text{then } 3(-1) + y = 6. \quad \text{Thus, } y = 9.$$
$$\text{If } x = 0, \quad \text{then } 3(0) + y = 6. \quad \text{Thus, } y = 6.$$
$$\text{If } x = 1, \quad \text{then } 3(1) + y = 6. \quad \text{Thus, } y = 3.$$
$$\text{If } x = 2, \quad \text{then } 3(2) + y = 6. \quad \text{Thus, } y = 0.$$
$$\text{If } x = 3, \quad \text{then } 3(3) + y = 6. \quad \text{Thus, } y = -3.$$

For each value of x that you choose, you will get a linear equation in y. When you solve that equation, you will get a number for y. A solution to $3x + y = 6$ will be a pair of numbers (value of x, value of y). The pair of numbers is ordered in that the value of x is always stated first, followed by the value of y.

A **solution** to a linear equation in two variables, x and y, is an ordered pair of numbers (a, b) such that when you substitute a for x, you get b as the solution to the resulting equation in y.

Is $(5, -12)$ a solution to $3x + y = 6$? To find out, replace x with 5 and solve for y.

$$3(5) + y = 6$$
$$15 + y = 6$$
$$y = -9$$

Since y does not equal -12, you may conclude that $(5, -12)$ is not a solution.

One thing is clear from the discussion above. If you have a linear equation in two variables, then whenever you choose a value of x, you will get a value of y. Therefore, there are infinitely many solutions to the equation. However, not every pair of numbers that exists will be a solution to the equation.

Using what you know about solving linear equations in one variable, you can write a formula for the entire solution to $3x + y = 6$. Let $x = a$, then $3a + y = 6$. Therefore, $y = -3a + 6$. The complete set of solutions to $3x + y = 6$ can be expressed by the ordered pair $(a, -3a + 6)$.

The Coordinate Plane

The equation $ax + by = c$, where a, b, and c are real numbers and neither a nor b is zero is called a *linear equation in two variables*. A *solution* to a linear equation in two variables x and y is any ordered pair of numbers (x, y) that makes the equation true.

• The term "linear" indicates that the variables are not raised to any power greater than 1.

• There is another reason for calling $ax + by = c$ a linear equation. If you graph all the ordered pairs that make the equation true, you will get a line.

When you place two identical number lines together so that they are perpendicular to one another and so that their 0 points coincide, the result is the **coordinate plane** as shown at right below.

• The number lines are called the *coordinate axes*, usually the **x-axis** and the **y-axis**. Their intersection is called the *origin*.

• Points in the coordinate plane correspond to ordered pairs (x, y) of real numbers. The first member of the ordered pair is called the *x-coordinate* and the second member of the ordered pair is called the *y-coordinate*.

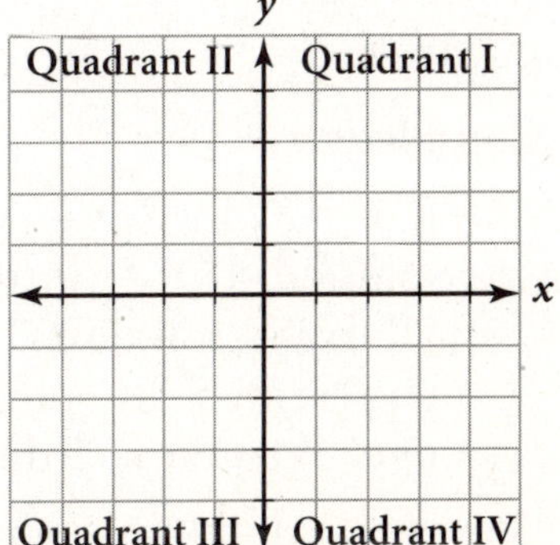

variable: A letter that is used to represent numbers in an algebraic expression.

linear equation: An equation whose graph is a line.

solution: The values of x and y that make an equation true.

coordinate plane: A plane that is divided into four regions by a horizontal and a vertical number line.

x-axis: The horizontal number line in a coordinate plane.

y-axis: The vertical number line in a coordinate plane.

quadrant: One of the four regions into which a horizontal and vertical line divide a plane.

- The coordinate axes divide the plane into four **quadrants.** Points on the coordinate axes are not in any quadrant.

The coordinate plane is used to represent ordered pairs of numbers graphically.

Consider $(3, 4)$ and $(-1, -2)$.

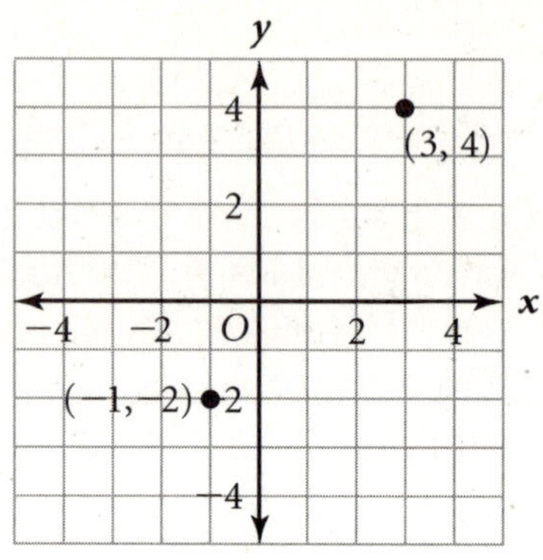

For any given ordered pair, use the horizontal number line to locate the first number and the vertical number line to locate the second number in the pair. For example, to graph $(3, 4)$, count 3 units to the right from O and then count 4 units up. For $(-1, -2)$, count 1 unit to the left of O and then count 2 units down.

ordered pair: The x- and y-coordinates that give the location of a point in a coordinate plane, indicated by two numbers in parentheses, (x, y).

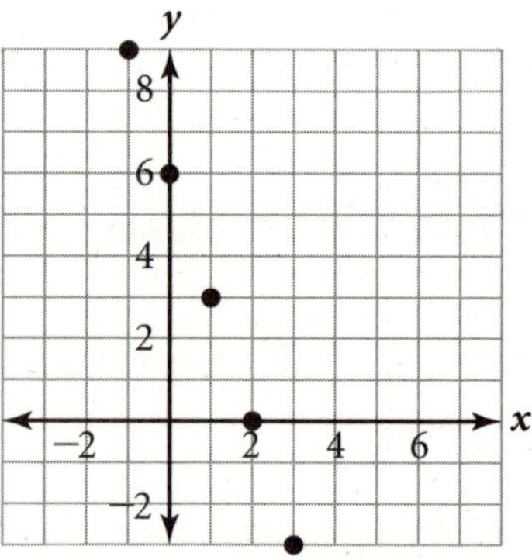

The diagram at right shows the graphs of the **ordered pairs** $(-1, 9)$, $(0, 6)$, $(1, 3)$, $(2, 0)$ and $(3, -3)$, which are some of the solutions to $3x + y = 6$. Notice that the points seem to show a pattern. If you take a ruler and place it over the diagram, you will see that the edge of the ruler goes through each of the points. That is, the points all lie along a definite line in the coordinate plane. This suggests that the graph of $3x + y = 6$ is a line.

An equation in two variables is an algebraic object. One of the first problems that students need to solve in algebra is how to make a geometric representation of an equation.

$$\text{algebraic object} \longrightarrow \text{geometric object}$$
$$\text{equation} \longrightarrow \text{graph}$$

Given a linear equation in two variables, how can I represent this equation graphically?

Consider $2x + 3y = 6$.

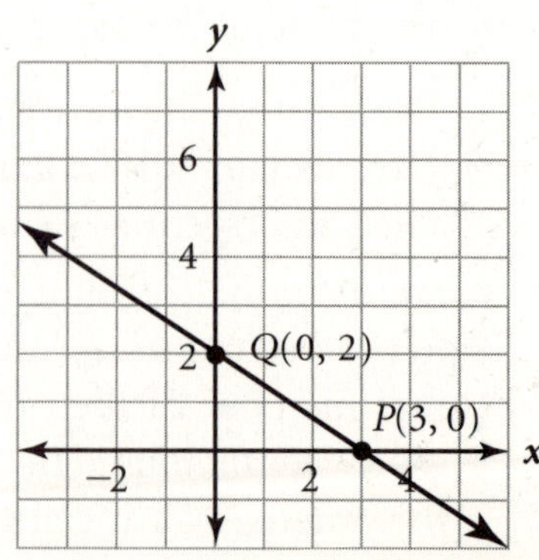

If $x = 0$, then $y = 2$. Thus, $(0, 2)$ is a solution. If $y = 0$, then $x = 3$. Thus, $(3, 0)$ is a solution. These solutions are represented by points $P(3, 0)$ and $Q(0, 2)$ on the coordinate plane at right.

The line containing points $P(3, 0)$ and $Q(0, 2)$ is the graph of $2x + 3y = 6$.

The Graph of a Line

- Two points in the plane determine exactly one line.

- If two ordered pairs are solutions to a given linear equation, then the two ordered pairs can be used to sketch the graph of the equation.

- When $x = 0$, the equation can be solved to find the corresponding value of y. When $y = 0$, the equation can be solved to find x. These two ordered pairs, $(0, y)$ and $(x, 0)$, are the points where the line crosses each axis.

- The two ordered pairs just found can be graphed. The graph of the equation will be completely determined by drawing a line passing through the two points.

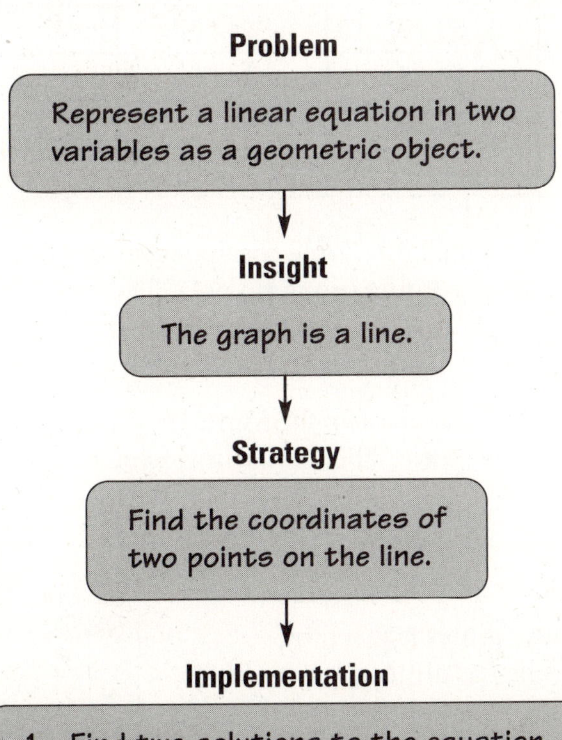

Applications of Linear Equations in Two Variables

Lauren has \$5.00 in nickels and dimes. How many of each might she have?

You can answer this question by thinking about at least some of the possible combinations of nickels and dimes. For example, if she has no nickels, then she must have 50 dimes. Other combinations are shown in the table below.

number of nickels	value of nickels	remaining value	number of dimes
1	\$0.05	\$4.95	not a whole number
2	\$0.10	\$4.90	49
3	\$0.15	\$4.85	not a whole number
4	\$0.20	\$4.80	48
5	\$0.25	\$4.75	not a whole number
6	\$0.30	\$4.70	47
. . .	. . .	. . .	. . .

From the table, you can tell that some combinations of nickels and dimes are possible, but that other combinations are not. There must be a better way to find out what combinations of coins are possible and what combinations of coins are not possible. An equation in two variables can help.

Let n represent the number of nickels and let d represent the number of dimes that Lauren may have. Then you can reason as follows.

❶ The value of the nickels is given by
   $5 \times$ (number of nickels),
   where the value is in pennies.

❷ The value of the dimes is given by
   $10 \times$ (number of dimes),
   where the value is in pennies.

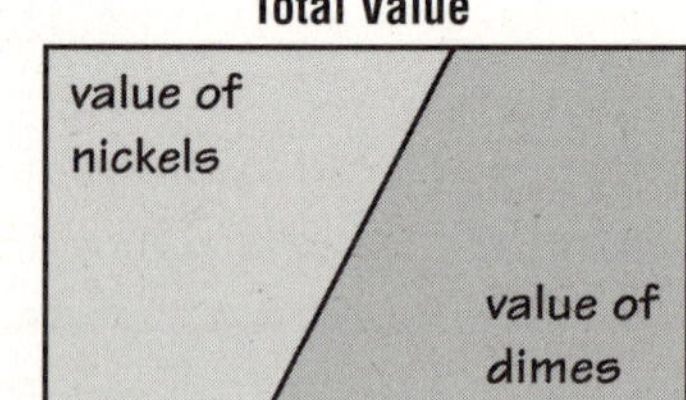

❸ The total value of the coins is
   500 pennies (\$5.00).

❹ If you add the value of the nickels and the value of the dimes, you should get 500, the total value (in pennies) of all the coins.

Now replace the phrases "value of nickels" and "value of dimes" with letters, or variables. Any letter can represent a variable. In this case, we choose n for "number of nickels" and d for "number of dimes."

Now we can write the equation.

$$5n + 10d = 500$$

A solution to the equation in this problem is an ordered pair of numbers (number of nickels, number of dimes), or (n, d). The diagram at right shows the graph of the solutions. Notice that the dashed line extends from $(0, 100)$ to $(50, 0)$. These points represent the combinations *no dimes and 100 nickels* and *50 dimes and no nickels*, respectively. The ordered pair $(30, 40)$ is on the dashed line segment because it is possible for Lauren to have 30 dimes and 40 nickels.

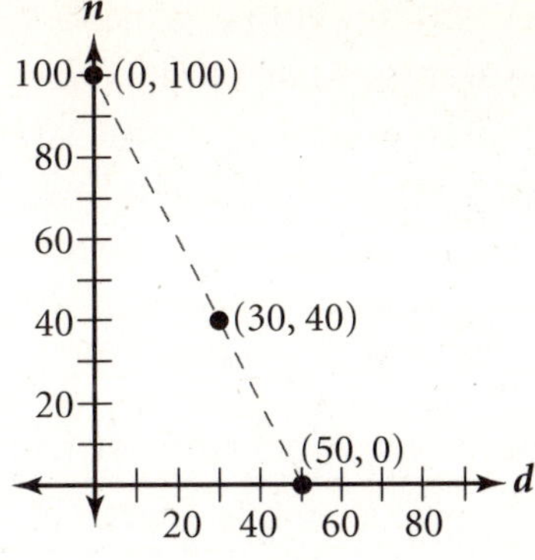

To help you better answer the question about coin combinations, you can rewrite the equation in another form. To get this form, the given equation can be transformed by using principles of equality until it looks like $n = -2d + 50$. This equation has the advantage that one variable is given in terms of the other. In this equation, n is given in terms of d.

$$5n + 10d = 500$$

$$n + 2d = 100 \qquad \textit{Divide each side by 5.}$$

$$2d = -n + 100 \qquad \textit{Subtract } n \textit{ from each side.}$$

$$d = -\frac{n}{2} + 50 \qquad \textit{Divide each side by 2.}$$

The equation $d = -\dfrac{n}{2} + 50$ will tell how many dimes Lauren has if she has a specified numbers of nickels. You can also tell from this equation that if n is an even counting number, then you will get a counting number for d. This occurs because in the equation, n is being divided by 2, an even number. For the same reason, if n is an odd counting number, you will not get a counting number for d. This explains why in the table on the preceding page, you did not get a whole number of dimes when you considered 1, 3, 5, etc. as numbers of nickels. The table below shows a partial listing of solutions.

n	0	2	4	6	8	10	12	. . .	100
d	50	49	48	47	46	45	44	. . .	0

Many problem-solving situations require two variables to solve a problem. The following problem, although quite different from the coin problem, requires the use of two variables.

A storage tank contains 300 gallons of water. A discharge valve is opened and water is drained out at the rate of 60 gallons per minute. How much water is in the tank after a specified number of minutes?

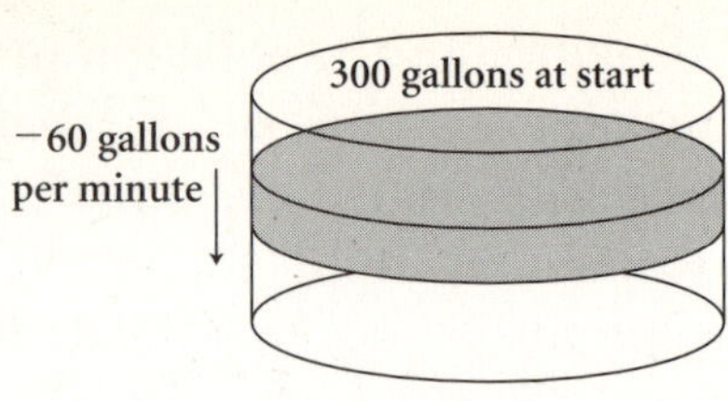

Using the diagram at right above to aid your reasoning, you will be able to write an equation that relates volume and elapsed time.

amount of water in the tank = initial volume − amount of water drained

The amount of water drained depends on the amount of time the value is opened.

amount of water drained = 60 × (number of minutes the valve is open)

Next we introduce variables to replace verbal phrases. Let t represent the number of minutes the value is opened. Let V represent the amount of water in the tank after t minutes.

amount of water in the tank = initial volume − amount of water drained

$$V = 300 - 60t, \text{ or}$$

$$V = -60t + 300$$

Notice that for a given value of t, you can calculate V, the remaining volume. The equations $5n + 10d = 500$ and $V = -60t + 300$ have characteristics in common.

- Each equation has two variables.
- The variables are each raised to the first power; that is, the variables do not have exponents other than 1.

At this point we can revisit the question of how many nickels and dimes Lauren might have. The graph on the preceding page shows the solutions to $5n + 10d = 500$. Notice that the line segment is contained in Quadrant I. This is because the number of nickels and the number of dimes must be 0 or more. Notice also that the line is not drawn with a solid line. This is because not every point on the line represents a solution. For example, $(0, 100)$, $(2, 49)$, and $(50, 0)$ are on the graph. There are no points on the graph for values of n that are odd. The graph is a *discrete* set of points.

We now return to the question of the tank being drained. The graph at right shows the relationship between volume and time. Time is plotted along the horizontal axis and volume is plotted along the vertical axis.

Notice that the graph relating V to t is also contained in Quadrant I. The graph ends at points on the positive x-axis and the positive y-axis. The graph is, however, *continuous*. That is, for each nonnegative value of t, there is a corresponding value of V.

Before completing this discussion of linear equations in two variables, it is worth mentioning a few things about linear equations and graphs.

> ### Linear Equations and Graphs
>
> - If there are no restrictions on variables, the graph of $ax + by = c$ is a line.
>
> - In application contexts, where the variables represent real quantities, there may be restrictions that make the graph a line segment contained in a quadrant, usually Quadrant I. In a particular problem modeled by a linear equation in two variables, the graph may be a finite set of points that lie along a line. In another real-world problem modeled by a linear equation in two variables, the graph may be a line segment, a portion of a line with two endpoints.

When reading application problems, you should be attentive to the meanings of the variables and any restrictions on them.

Slope and y-intercept

The problem involving the volume of water in the tank gave the equation below.

$$V \quad = \quad -60t \quad + \quad 300$$

volume at time t = (constant rate) $\times$ drainage time t + initial amount

Many application problems can be represented by this type of equation.

Consider this application problem.

> Initially a solution in a container has temperature 68°F. Its temperature is raised at the constant rate of 5°F per minute. Represent temperature, T, in degrees Fahrenheit as an equation in terms of elaped time, t, in minutes.

temperature at time t = (rate) $\times$ heating time t + initial temperature

$$T \quad = \quad 5t \quad + \quad 68$$

Notice that the initial temperature and rate of temperature increase are sufficient to determine the equation for T.

There are many application problems in which a rate of change and an initial state determine a linear equation in two variables. Therefore, you can make the generalization below.

> ### Constant Rate of Change and Initial State
>
> The equation $y = mx + b$ represents a linear equation in two variables. The constant rate of change is given by m and the initial state, the state when $x = 0$, is given by b.
>
> $$y = (\text{constant rate of change})x + (\text{initial state})$$

In Chapter 1, we said that one of the benefits of giving algebra students application problems is that these problems are an important link between the symbolic structure of algebra and the meaning represented by these symbols. An application problem provides a well-defined context and relevant information, which allows students to use background knowledge, thus making problem solving more meaningful. Unfortunately, not every introductory algebra course includes applications. For example, students may be given an equation like $y = \frac{1}{2}x + 1$. The number $\frac{1}{2}$, the coefficient of x, may have nothing to do with rate and the number 1 may have nothing to do with initial state. Rather than being concerned with application issues, students are asked to study the formal properties and the algebraic names of these numbers. However, helping students to develop an understanding of algebra concepts is as important as helping them learn the formal aspects of algebra. We now return to our equation. Consider the following question as you read the argument that follows.

Since the graph of $y = \frac{1}{2}x + 1$ is a line, what can be said about that line?

To help answer the question, consider the graph of $y = \frac{1}{2}x + 1$ shown on the next page.

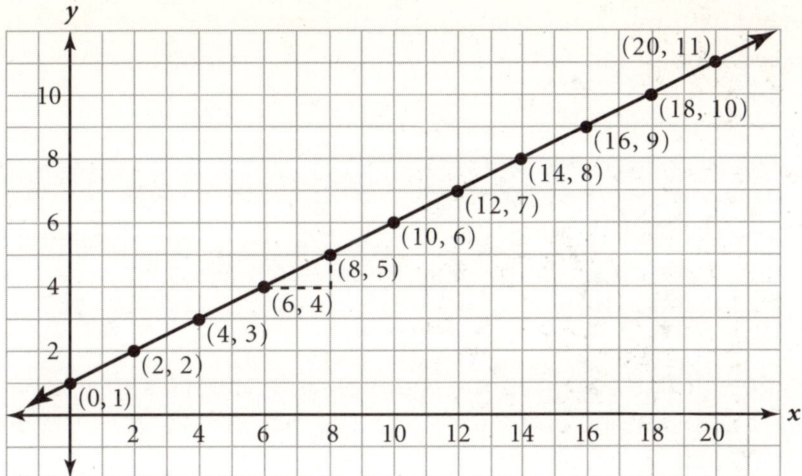

Notice that for each increase of 2 in x, there is an increase of 1 in y. This is illustrated for the ordered pairs $(6, 4)$ and $(8, 5)$. Some teachers might put a diagram like this one on the whiteboard and ask students to find the increase in y for various increases of x. Students will find that the ratio of the differences in y to the differences in x will always be 1 to 2. This ratio is exactly the coefficient of x, $\frac{1}{2}$.

After completing this exploration, students might ask a question like the following.

> Suppose that I pick any two points on the graph and calculate the ratio of the differences in y to the difference of the corresponding values of x. Will I always get the ratio 1 to 2?

Some student choices and calculations are shown below.

$(4, 3)$ and $(16, 9)$	$(2, 2)$ and $(14, 8)$	$(0, 1)$ and $(18, 10)$
$\dfrac{9 - 3}{16 - 4} = \dfrac{6}{12} = \dfrac{1}{2}$	$\dfrac{8 - 2}{14 - 2} = \dfrac{6}{12} = \dfrac{1}{2}$	$\dfrac{10 - 1}{18 - 0} = \dfrac{9}{18} = \dfrac{1}{2}$

All the choices and calculations give the same ratio: 1 to 2. This is a very important finding. Given a slant line in the coordinate plane, any pair of points on it will give the same ratio for the difference of two values of y to the difference of the corresponding values of x.

Slope

If $P(x_1, y_1)$ and $Q(x_2, y_2)$ are on the graph of $y = mx + b$, then $\dfrac{y_2 - y_1}{x_2 - x_1}$ will always be the same ratio. This ratio is called the *slope of the line* and is represented by m.

slope: A measure of the steepness of a line. In the equation $y = mx + b$, m denotes the slope of the line.

y-intercept: The *y*-coordinate of the point where a line crosses the *y*-axis. In the equation $y = mx + b$, *b* denotes the *y*-intercept.

The second defining characteristic of a line is where it crosses the *y*-axis. The **y-intercept** of a graph is the *y*-coordinate of the point where the graph crosses the *y*-axis.

> ### Slope and y-intercept of a Line
>
> The graph of $y = mx + b$ is a straight line in the coordinate plane. The line has slope given by *m* and *y*-intercept given by *b*.

For example, if you are given the equation $y = 1.5x - 4$, you can say that the graph is a straight line with slope 1.5 that crosses the *y*-axis 4 units below the *x*-axis.

Give students enough practice and time to understand the concepts of slope and *y*-intercept. Use as many application problems as necessary. Remember, you will know when students have had enough practice when they can transfer their knowledge from one situation to another with little or no difficulty. Students can create a brief summary such as the one below to help them make the connection between symbols and meaning.

$$y = mx + b$$

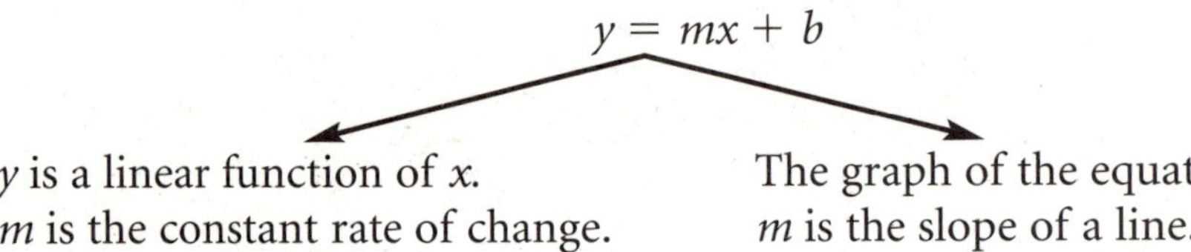

y is a linear function of *x*.
m is the constant rate of change.
b is the initial state.

The graph of the equation is a line.
m is the slope of a line.
b is the *y*-intercept of the graph.

Finding an Equation for a Line

Geometry and algebra are interrelated disciplines. Frequently, geometry problems can be solved by using algebra. Consider the problem below.

> Find a representation of all points on the line that passes through $P(-2, 3)$ and $Q(5, -7)$.

This is a geometry problem because it involves points and lines. You could graph $P(-2, 3)$ and $Q(5, 3)$, draw the line through them, and present the picture as the representation. However, this representation does not tell anything about the points between $P(-2, 3)$ and $Q(5, 3)$.

Consider the algebraic reasoning below.

> Every line in the plane can be represented by a linear equation. Since the two given points determine a line with some slope, maybe I can

use slope to help find an algebraic representation of all the points on the line.

This is a problem-solving strategy. Its step-by-step implementation is shown below.

❶ Find the slope of the line containing $P(-2, 3)$ and $Q(5, -7)$.

$$\frac{-7 - 3}{5 - (-2)} = \frac{-10}{7} = -\frac{10}{7}$$

❷ Let $A(x, y)$ represent any point on the line and use slope again.

$$\frac{y - 3}{x - (-2)} = -\frac{10}{7}$$

❸ Solve the equation above for y in terms of x.

$$\frac{y - 3}{x + 2} = -\frac{10}{7}$$

$$y - 3 = -\frac{10}{7}(x + 2)$$

$$y = -\frac{10}{7}(x + 2) + 3$$

$$y = -\frac{10}{7}x - \frac{20}{7} + 3$$

$$y = -\frac{10}{7}x + \frac{1}{7}$$

The equation $y = -\frac{10}{7}x + \frac{1}{7}$ tells you that for each value of x you choose, the y-coordinate of any point on the line is found by calculating $-\frac{10}{7}x + \frac{1}{7}$.

Reflection on the discussion above suggests the following.

Using Multiple Resources to Solve Problems

- Students can use algebra to find the solution to a geometry problem.

- The concept of the slope of a line is often critical to solving problems involving lines.

- Problem-solving strategies are needed even to solve problems that appear to be straightforward or routine.

These reflections are not insignificant. Indeed, a central lesson for students to learn is that algebra is not just a matter of solving equations or manipulating symbols. Rather algebra is a problem-solving tool. Algebraic methods, concepts, and calculations are integral to solving problems that may not be solvable otherwise.

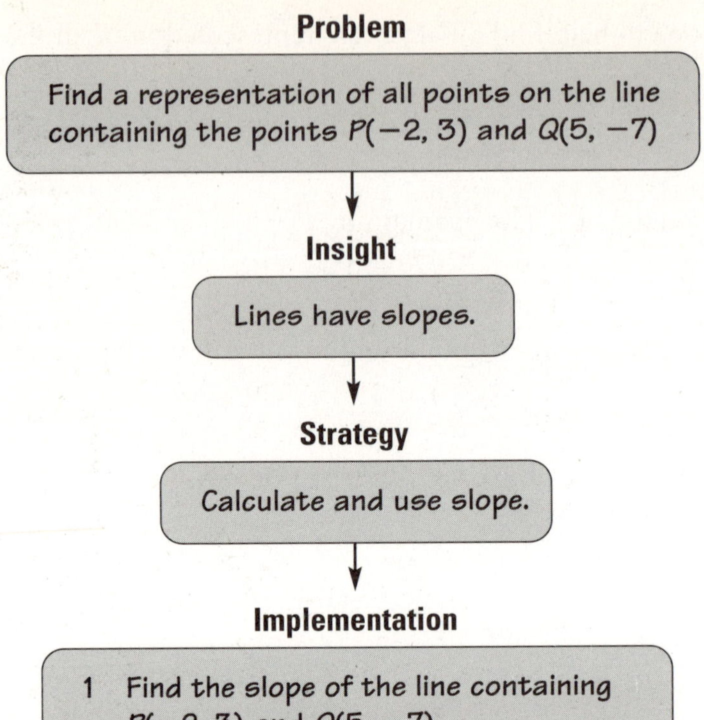

Students will also be asked to find an equation of a line given the coordinates of one point on it and its slope. This type of problem is even simpler to solve than the problem involving coordinates of two points since the slope is already given. The following equation does not involve numerical information but rather involves a description of what is given in a problem.

$$\frac{y - \text{given } y\text{-coordinate}}{x - \text{given } x\text{-coordinate}} = \text{given slope}$$

Using the equation above, you can find a formula for y.

$$y - \text{given } y\text{-coordinate} = (\text{given slope})(x - \text{given } x\text{-coordinate})$$

$$y = (\text{given slope})(x - \text{given } x\text{-coordinate}) + \text{given } y\text{-coordinate}$$

For example, if a line contains $P(3, 7)$ and has slope 2, you can write the following equation for y.

$$y = 2(x - 3) + 7, \text{ or } y = 2x + 1$$

In Chapter 1 we explained how algebra skills relate to one another. For this reason, students need to learn skills in a sequential order. Skills previously learned, together with current strategies and concepts, are brought to bear on current problem solving. If students do not successfully learn skills as they progress in the study of algebra, they will find subsequent problem solving increasingly difficult and frustrating.

Solutions and Graphs of Linear Inequalities in Two Variables

A **linear inequality in two variables** is any inequality that can be written in one of these forms.

$$ax + by < c \qquad ax + by \leq c \qquad ax + by > c \qquad ax + by \geq c$$

The inequality $x - 2y < -6$ is an example of a linear inequality in two variables.

linear inequality: An inequality whose form is like that of a linear equation with the equal sign replaced by an inequality symbol.

Most algebra textbooks do not explain in detail why the solution to a linear inequality in two variables is a *half plane*. Rather, the statement is made that the solution is a half plane and students are almost immediately taken into the process of finding that half plane. In the following discussion, you will have an opportunity to look at the reasons that explain why the solution to a linear inequality in two variables is a half plane.

Consider $x - 2y < -6$.

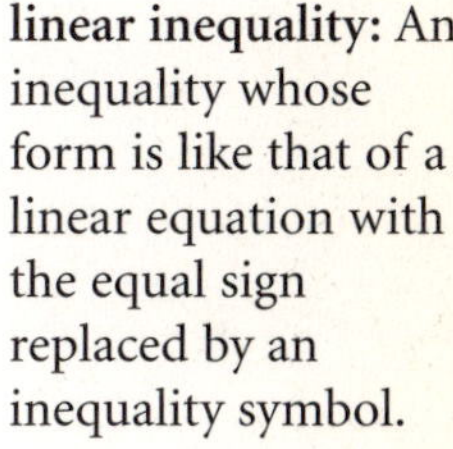

Write y in terms of x. $y > \frac{1}{2}x + 3$

If $x = 0$, $y > 3$, the open ray pointing up from 3.

If $x = 2$, $y > 4$, the open ray pointing up from 4.

If $x = 4$, $y > 5$, the open ray pointing up from 5.

If $x = 6$, $y > 6$, the open ray pointing up from 6.

If $x = 8$, $y > 7$, the open ray pointing up from 7.

If $x = 10$, $y > 8$, the open ray pointing up from 8.

According to these results, for each value of x, there is a ray, $y > \frac{1}{2}x + 3$, that is part of the solution to $x - 2y < -6$. The solution to $x - 2y < -6$ is the infinite collection of rays obtained from $y > \frac{1}{2}x + 3$.

These rays fill up a part of the coordinate plane that is above the line passing through the open circles.

The diagram contains important information. For example, the line through the open circles has a linear equation associated with it. Also, points in the region above the line make the inequality true, and points below the line make the inequality false.

The reasoning above suggests a method for graphing a linear inequality in two variables.

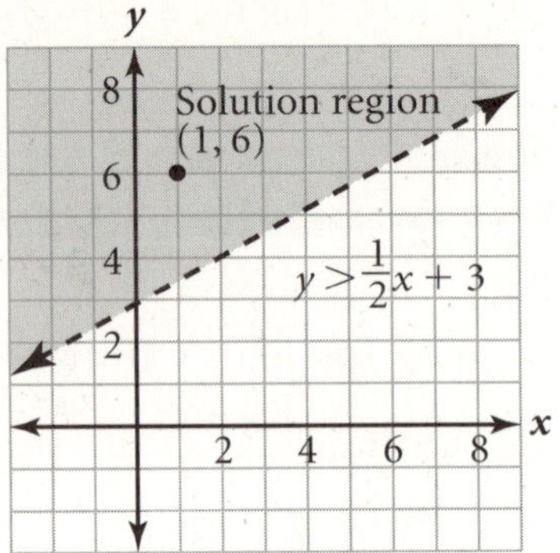

❶ Sketch the line that is the graph of the associated linear equation in two variables.

❷ Decide whether the solution rays point up from the line or down from it. For $x - 2y < -6$, for example, test $(1, 6)$ to see if it is a solution. Since $(1, 6)$ is a solution to $x - 2y < -6$, the region above the line is the solution.

❸ Decide whether the endpoints of the rays are included (solid endpoints) or excluded (open endpoints). In the case of $x - 2y < -6$, the endpoints are not included. Thus, the boundary of the solution region should be drawn with a dashed line.

Mathematical thinking used to solve a particular type of problem can often be summarized in such a way that it can become a procedure. Students will be amazed to learn that they can solve all linear inequalities in two variables with the aid of the following decision chart. You can show this chart on the whiteboard.

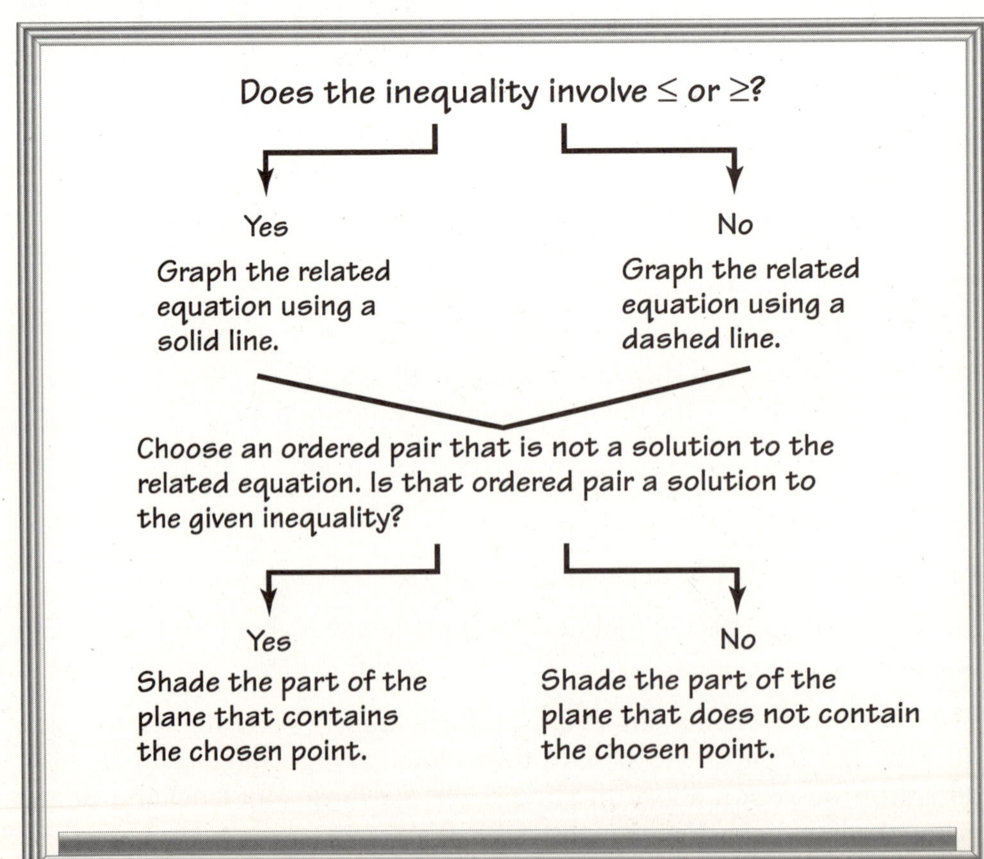

Consider how the decision chart on the previous page is used to graph $2x + 5y < 20$ and $3x - 4y \geq 7$.

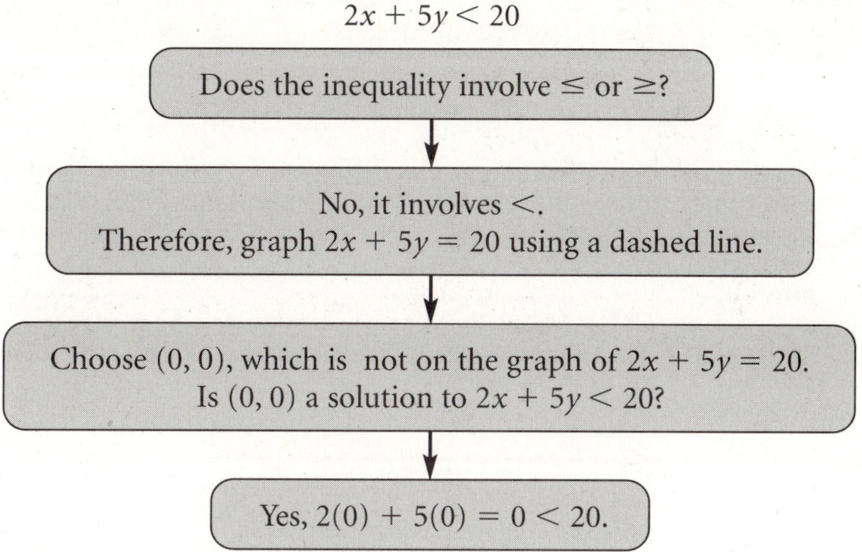

The graph of $2x + 5y < 20$ is the part of the coordinate plane bounded by the dashed graph of $2x + 3y = 20$ and containing $(0, 0)$.

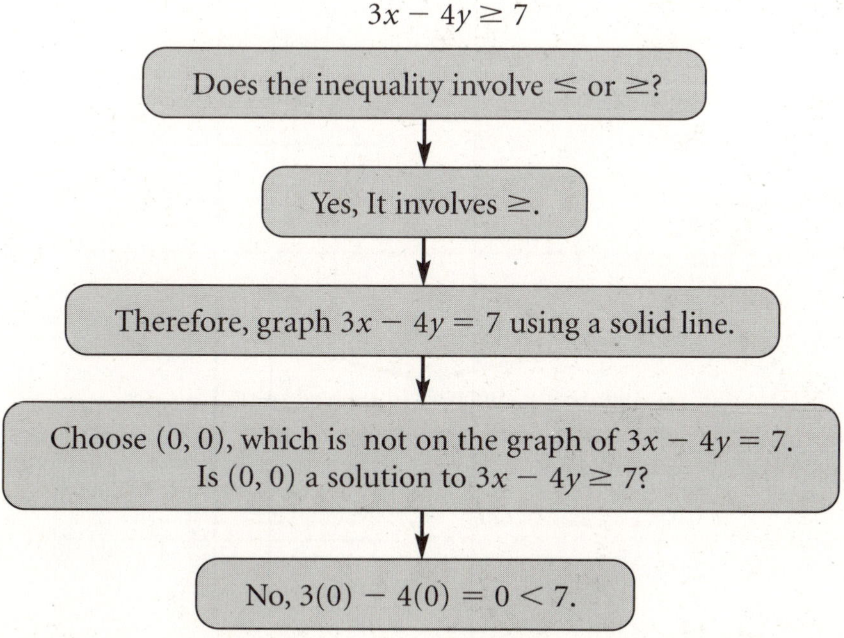

The graph of $3x - 4y \geq 7$ is the part of the coordinate plane bounded by the solid graph of $3x - 4y \geq 7$ and not containing $(0, 0)$.

Students can create their own decision chart to guide their work with linear inequalities in two variables. They might be surprised to learn that their thinking can be organized in such a compact way. This gives students the opportunity to become independent thinkers and decision makers.

A linear equation or inequality may contain only one variable. The solution to such an equation or inequality depends on whether it occurs in one dimension or in two dimensions. In one dimension, solutions are points or rays. In two dimensions, solutions are lines or half planes. The chart below illustrates this important distinction.

One Dimension	**Number Line**
$x = 2$	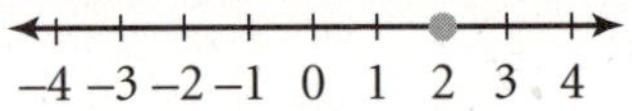
$x \geq 2$	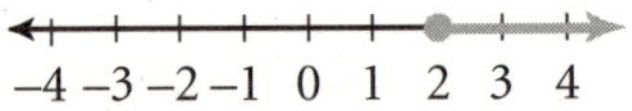
Two Dimensions	**Coordinate Plane**
$x = 2$	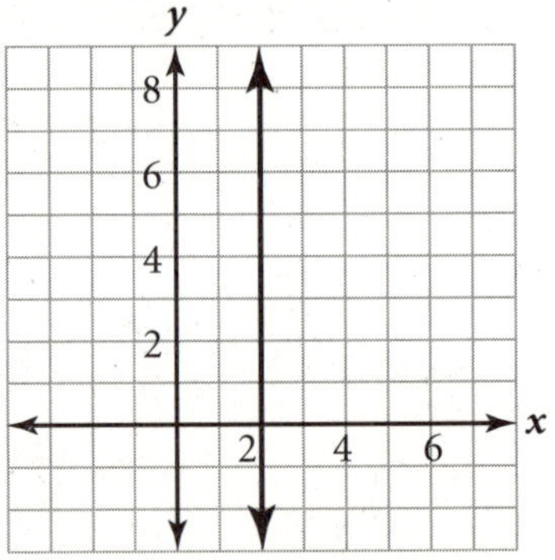
$x \geq 2$	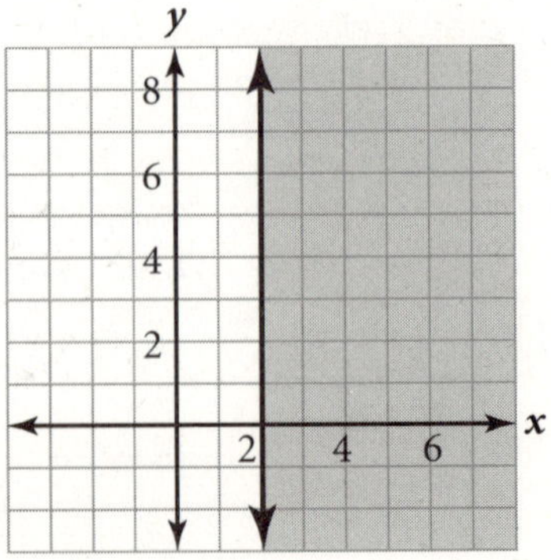

Keeping an Algebra Journal

Both algebra teachers and algebra students will find that keeping records in the form of a journal will be very helpful to summarize what has been learned, such as major ideas, methods, concepts, and vocabulary terms. Students can refer to journal notes to study before quizzes and tests, to complete homework assignments, and to remember what questions to ask in class.

Sample notes are shown on this journal page. Notice that the journal entries are important reminders for the student to practice certain kinds of problems and to participate actively in class by asking questions.

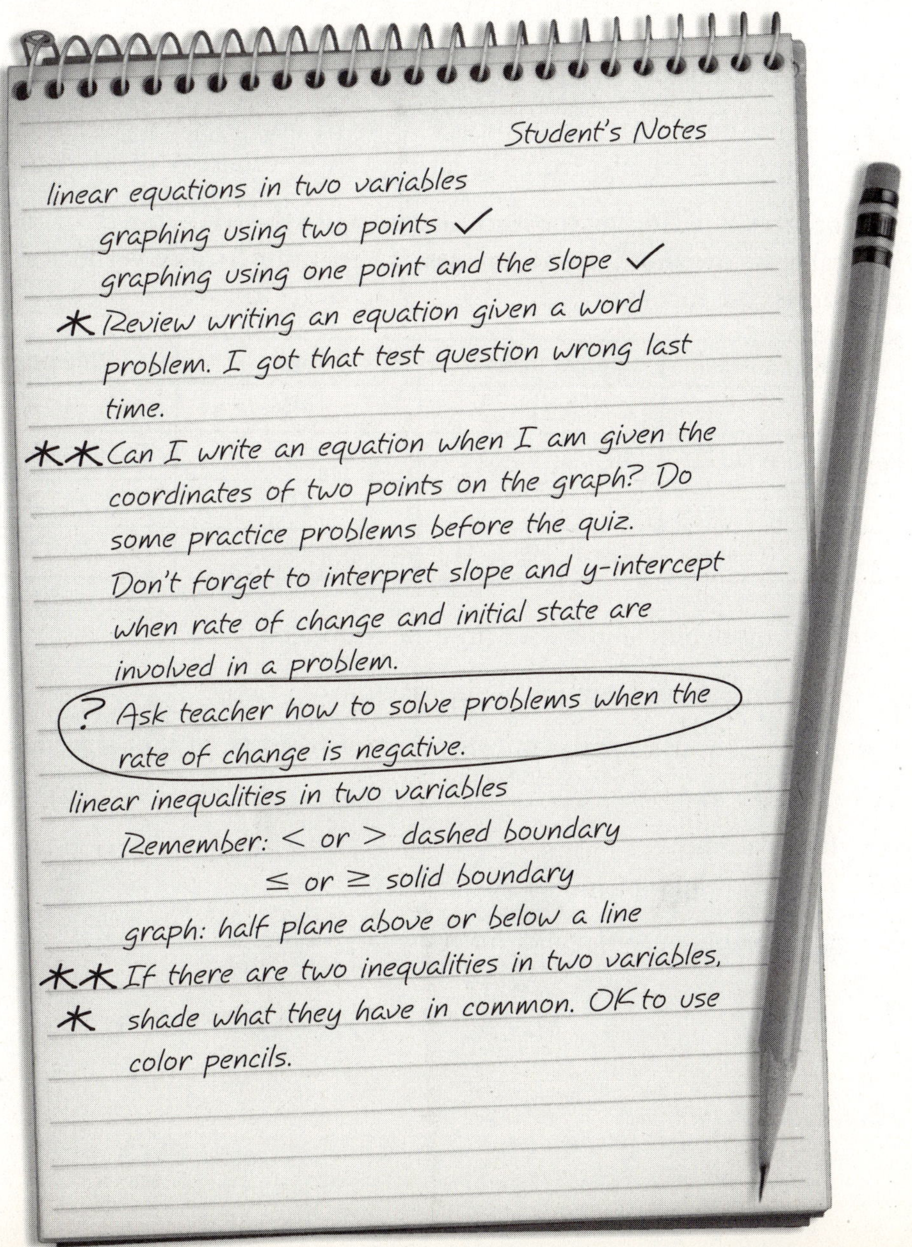

SYSTEMS OF LINEAR EQUATIONS AND INEQUALITIES

Systems of Linear Equations

Consider the number problem below.

> I am thinking of two numbers. When I add them, I get 19. When I subtract the smaller from the larger, I get 5. What are the the numbers?

One approach to finding the numbers is to use the trial-and-error method. The table below shows some guesses at the numbers and verification that the guesses are incorrect.

Smaller number	Larger number	Sum	Difference
8	11	19	3 ✘
6	13	19	7 ✘
−2	21	19	23 ✘
0	19	19	19 ✘

Is there a more efficient way to find the numbers?

We can think about a graphical approach.

Let m represent the smaller number and let n represent the larger number.

m = smaller number

n = larger number

When you add them, you get 19.

When you subtract the smaller from the larger, you get 5.

$$m + n = 19$$
$$n - m = 5$$

Graph each of the equations on the same coordinate plane. Then look at the graphs to see if the lines intersect. If they do, the intersection point will give the pair of numbers, m and n, that satisfy both equations.

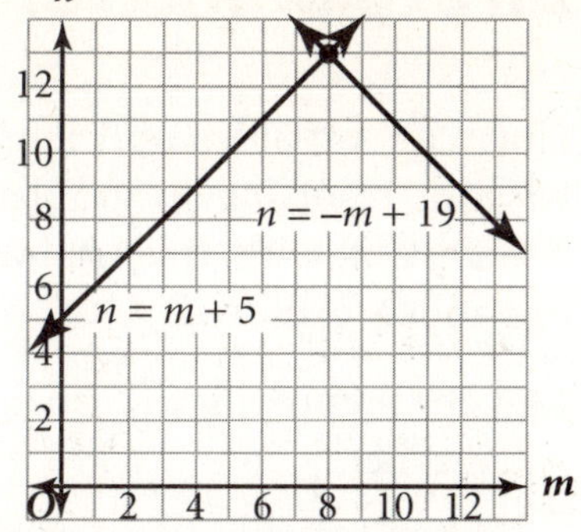

$$m + n = 19 \longrightarrow n = -m + 19$$
$$n - m = 5 \longrightarrow n = m + 5$$

The graphs are shown at right. The lines do intersect. When you locate the point of intersection, you will find that its coordinates are (7, 12). This pair of numbers gives the correct values of m and n, respectively.

$$7 + 12 = 19 \text{ and } 12 - 7 = 5 \text{ ✔}$$

The graphical approach worked. The numbers are 7 and 12.

From this discussion, you can see that an algebraic problem can be solved with the aid of coordinate geometry. This method is obviously more efficient than the trial-and-error method used earlier. The graphical approach may seem cumbersome and a bit time consuming, but at least it is reliable in that it delivers the solution.

Some pedagogical reasons for students to explore **systems of linear equations** by graphing include:

system of equations: Two or more equations in two or more variables.

- Students see another connection between algebra and geometry. Although the problem is stated in words and represented algebraically, it is a graphical approach that delivers the solution.

- Students can be given other systems of equations and discover that the graphical approach has limitations. For example, when students graph $x + y = 6$ and $y = \frac{2}{3}x + 2$, they will find that the lines

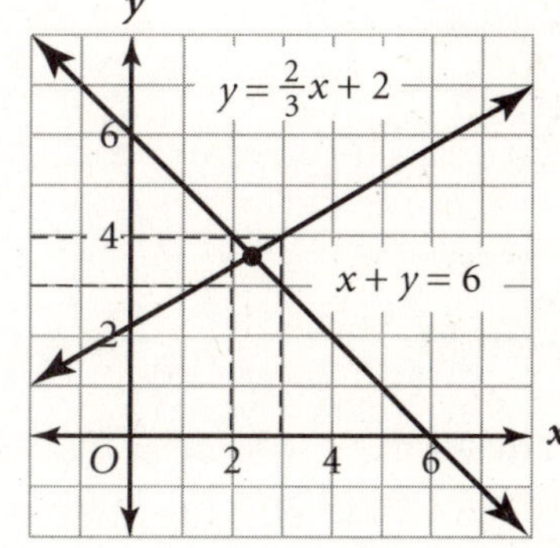

intersect but not in a point whose coordinates are integers. For this pair of equations, students can say that $2 < x < 3$ and $3 < y < 4$ but they cannot determine the exact values of x and y. Thus, students can begin to appreciate the need for algebraic methods that replace the graphical approach.

An Exploration of Solution Methods

How can variables and equations be used to solve the number problem without resorting to graphs? Consider the method explained below and compare it with the trial-and-error method and the graphical method previously explored.

❶ Let m represent the smaller number and let n represent the larger number.

When you add them, you get 19. When you subtract the smaller from the larger, you get 5.

$$m + n = 19$$
$$n - m = 5$$

❷ If I can remove, or eliminate, one of the variables and obtain an equation in only one variable, then I can use what I learned about solving linear equations in one variable to find one of the numbers.

❸ If I can succeed in Step ❷, then I should be able to find the other number by subtracting the number found in Step ❷ from 19.

In the method below, Steps ❶, ❷, and ❸ are carried out. The arrows indicate how one step leads to the next one.

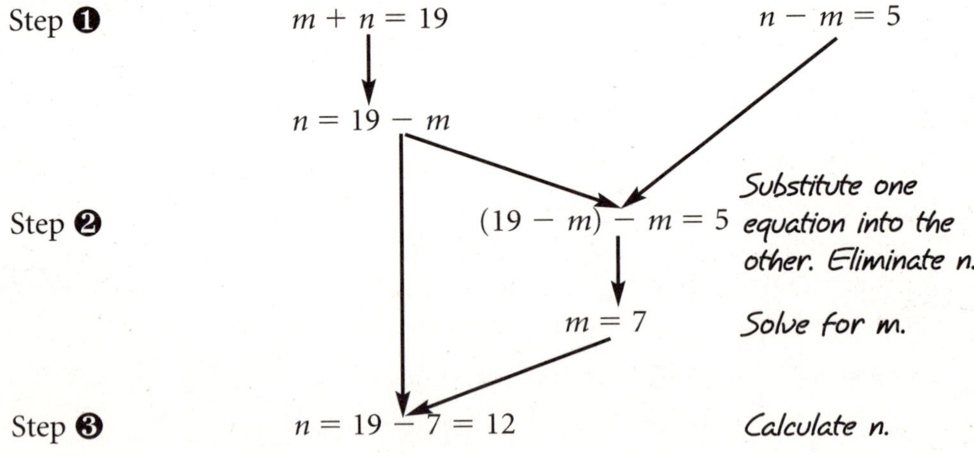

The strategy worked. The numbers are 7 and 12.
$7 + 12 = 19$ and $12 - 7 = 5$ ✔

substitution method: A method used to solve a system of equations in which variables are replaced with known values or algebraic expressions.

Because substitution of an expression for one variable is used as a crucial part of the solution method, this method is called the **substitution method.**

The following example also illustrates the use of the substitution method. We suggest to diagram the steps of this method and illustrate each step with a specific example. This will help students to organize the steps of the method and refer to the diagram as they complete their assignments.

Consider $\begin{cases} 3u + 2v = 12 \\ u - 2v = 3 \end{cases}$.

If I can solve one of the two given equations for one variable, in this case u, then I can make a new equation that contains only one variable, in this case v. Then I can solve the equation for v by using previously learned skills. This part of the method is shown as Part I below.

$$\begin{cases} 3u + 2v = 12 \\ \circledu - 2v = 3 \end{cases}$$

$\uparrow$

coefficient 1

Part I of the solution

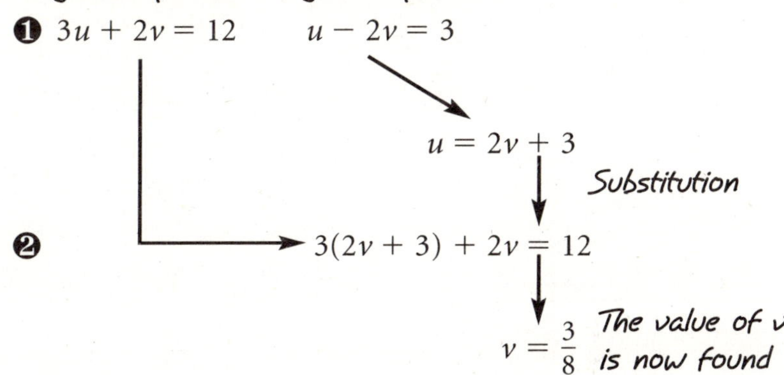

Once I have found a number for v, I can substitute that number in either equation to find u. This is shown as Part II below.

Part II of the solution

When these two major parts of the method are completed, the solution is found. That solution is $\left(\dfrac{15}{4}, \dfrac{3}{8}\right)$.

Notice in the solution above that the word *substitution* is used in two different steps. In Part I, the expression $2v + 3$ is substituted for the variable u. In Part II, the number $\frac{3}{8}$ is substituted for the variable v.

One of the reasons that students in an introductory algebra course study systems of equations in two variables is to solve application problems. After all, real-world problems are characterized by multiple variables that are usually interrelated.

Consider this problem. A technician has two solutions, A and B, that are to be mixed to make a

> solution A: 10% sugar
> solution B: 25% sugar
> solution C: 500 milliliters at 20% sugar

third solution, C. The data in the box above show the three solutions and their corresponding sugar content. How much of each solution, to the nearest whole number of milliliters, is needed to make the specified solution?

❶ You can organize the data in a table.

	A	B	C
amount of solution	a	b	500
amount of sugar	$0.10a$	$0.25b$	$(0.20)500$

❷ Write and solve a system of equations.

$$\begin{cases} a + b = 500 \\ 0.10a + 0.25b = 0.20(500) \end{cases} \longrightarrow \begin{cases} a + b = 500 \\ 2a + 5b = 2000 \end{cases}$$

Notice that the second equation was transformed into an equivalent equation by multiplying both sides by 20. When you solve the system by using the substitution method, you find that $a = 166\frac{2}{3}$ and $b = 333\frac{1}{3}$.

❸ Answer the question. The technician should use 167 milliliters of solution A and 333 milliliters of solution B.

Mixture problems are usually included in an introductory algebra course. In the solution above, a system of equations was used to solve a mixture problem. However, you might also notice that the problem could be solved by a single equation in one variable. A solution method using only one variable is shown below.

Let a represent the amount of solution A that is needed. Then the amount of solution B that is needed is $500 - a$. This is true because when you add the amount of solution A and the amount of solution B, you should get 500 milliliters. In symbols, $a + (500 - a) = 500$. Students then solve the equation below.

$$0.10a + 0.25(500 - a) = 0.20(500)$$

The solution to this equation is $166\frac{2}{3}$. Then students calculate $500 - 166\frac{2}{3}$ to find the amount of solution B. Notice that the equation above is exactly the same as the equation you get if you replace b with $500 - a$ in the system above.

Our discussion tells us that various solution methods give the same correct answer. This is important for both students and teachers in that students can find certain ways of representing and solving problems easier than some other ways, and teachers can observe and assess individual thinking processes.

You must have noticed that our discussion has paid a great deal of attention to the substitution method for solving a system of two linear equations in two variables. The following two reasons suggest why.

• The substitution method can always be used to solve a system of two linear equations in two variables.

• After substitution, you can apply skills previously learned to continue solving the problem.

The chapter on systems of equations in the pupil's edition presents various methods for solving a system of equations, including the substitution method. The elimination method and the multiplication-addition method are methods of solution that students should learn. With more than one method available, students need to know which method works best for a particular system of equation. Some guidelines are offered next.

Consider the two systems of equations below.

$$\begin{cases} m + n = 19 \\ n - m = 5 \end{cases} \qquad \begin{cases} 2a - 5b = -9 \\ -3a + 2b = 8 \end{cases}$$

In the system at right above, none of the coefficients of the variables is 1. Because of this, it is not easy to apply the substitution method to find the solution. This does not mean that the substitution method cannot be applied. However, the multiplication-addition method is designed to provide a more efficient method of solution.

Consider again $\begin{cases} 2a - 5b = -9 \\ -3a + 2b = 8 \end{cases}$.

Multiply each side of the first equation by 3 and the second equation by 2.

$$\begin{cases} 2a - 5b = -9 \\ -3a + 2b = 8 \end{cases} \rightarrow \begin{cases} 3(2a - 5b) = 3(-9) \\ 2(-3a + 2b) = 2(8) \end{cases} \rightarrow \begin{cases} \boxed{6a} - 15b = -27 \\ \boxed{-6a} + 4b = 16 \end{cases}$$

What results is a new system of equivalent equations. The solutions to the new system are the same as the solutions to the given system. Notice, however, that by a deliberate choice of multipliers the two terms involving a are now opposites of one another. This is the hallmark of the multiplication-addition method.

Now you can add the corresponding sides of the two equations. This addition is shown at right.

$$\begin{array}{r} 6a - 15b = -27 \\ + \; -6a + 4b = 16 \\ \hline -11b = -11 \end{array}$$

The equation that results is then solved.

$$-11b = -11 \rightarrow b = 1$$

If $b = 1$, then $2a - 5(1) = -9$. Therefore, $a = -2$.

This gives $(-2, 1)$ as the simultaneous solution to both equations.

In this discussion of systems of equations, you have read about three different ways to solve a system of equations. At this point, we recommend reviewing what has been studied so far. Information that is well organized will be easier to remember than information that is disorganized.

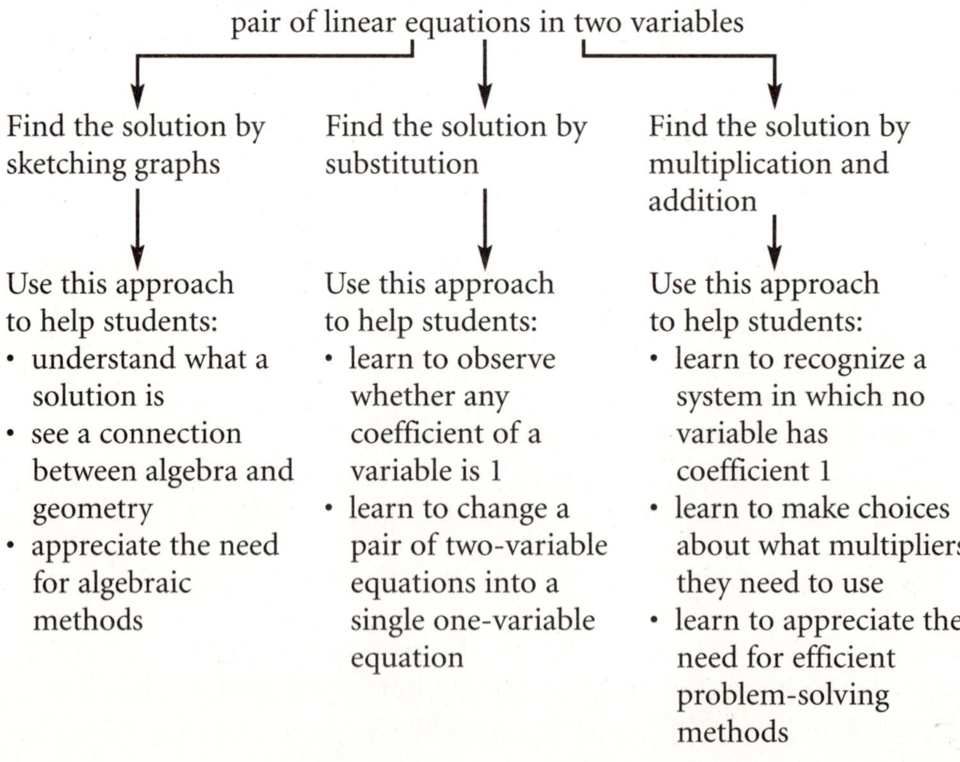

Learning the various skills and solution methods associated with systems of equations may require considerable time. We recommend to approach this topic in a gradual manner. Sufficient time should be allowed in lesson planning to make sure that students understand how solution methods work and why certain solution methods are more suitable for certain systems of equations than for others. If the lessons in a textbook are covered too quickly, then students may gain some procedural expertise but not reach the level of understanding required to sort all of this out.

To assess students' understanding of systems of equations and solution methods, you can give students a short quiz to find out what choices students made and why they made these choices.

Solution Scenarios

In addition to learning how to find a solution to a system, students should be taught that various solution scenarios may occur. For example, students should know that not every problem they are given has exactly one definite answer. As students explore this idea, you can explain to them that once a possibility for an answer occurs, that possibility in conjunction with the context of the problem can help explain the meaning of a particular answer. Consider the three different solution scenarios below.

$$\text{no solution} \qquad \text{exactly one solution} \qquad \text{infinitely many solutions}$$

Consider the three systems below.

$$A \begin{cases} x + y = 5 \\ x + y = 6 \end{cases} \qquad B \begin{cases} x + y = 5 \\ x - y = 3 \end{cases} \qquad C \begin{cases} x + y = 5 \\ 2x + 2y = 10 \end{cases}$$

- Observe and reflect on what you see in system A. How can $x + y$ equal 5 and, at the same time, $x + y$ equal 6? Our logic tells us that this is impossible. Therefore, system A has no solution.
- In system B, maybe $x + y$ can equal 5 and, at the same time, $x - y$ equal 3. When you apply a solution method, you will find that $x = 4$ and $y = 1$. Therefore, system B has exactly one solution.
- In system C, the second equation, $2x + 2y = 10$, is nothing but the first equation multiplied by 2. This system of equations has an entire line as its solution. Therefore, system C has infinitely many solutions.

Because making careful observations is a valuable mathematical skill, a classroom discussion of the three systems above may help students become more alert to the existence of various solution possibilities. You may want to write the three systems on the whiteboard and ask the question "How are these systems alike and how are they different from one another?" Also give students some time to reflect and make observations, and then call on individual students for their observations. Students may begin to discover for themselves that not all systems of equations are alike. To confirm their own observations and conclusions, have students use the graphical approach to see what the solutions look like in a graph.

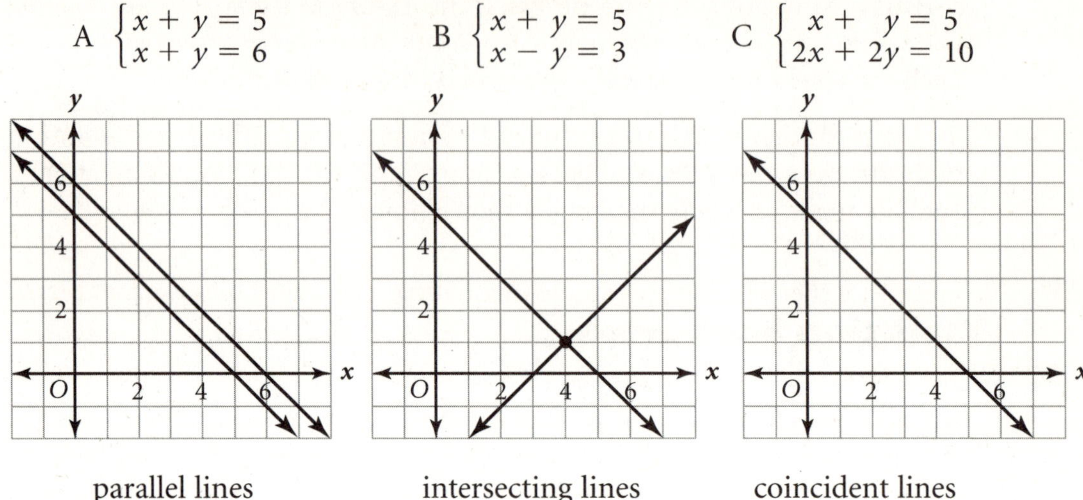

parallel lines   intersecting lines   coincident lines

In this section we have seen that a system of equations can have any one of three solution scenarios. The following decision chart is meant to help students decide whether a system of equations has one solution, no solution, or infinitely many solutions.

Many students will develop a high level of motivation to learn when they discover that what they are learning makes sense and that they can succeed at learning it. The topics of every chapter in the student book should be presented in a way that ensures that mathematical content is learned not only as a set of procedures to be mastered (procedural knowledge), but as meaningful concepts to explore in order to enhance students' understanding. The schematic representations and decision charts in this book are intended to help students organize knowledge in a way that facilitates future retrieval and application.

Systems of Linear Inequalities

Students can transfer much of what they have learned about systems of linear equations in two variables into the study of **systems of linear inequalities** in two variables. You can think of a system of linear inequalities as a system of linear equations with each equality sign ($=$) replaced by one of the four inequality symbols ($<$, $\le$, $>$, and $\ge$). An example of a system of linear inequalities in two variables is shown at right above.

$$\begin{cases} 2x - 5y \le 0 \\ y \le -\frac{1}{4}x - 2 \end{cases}$$

system of inequalities: Two or more inequalities in two or more variables.

Recall that the solution to a single inequality in two variables is a half plane. The solution for $y \ge \frac{2}{5}x$ is shown at left below. The solution for $y \le -\frac{1}{4}x - 2$, is shown at right below.

$$y \ge \frac{2}{5}x \qquad\qquad y \le -\frac{1}{4}x - 2$$

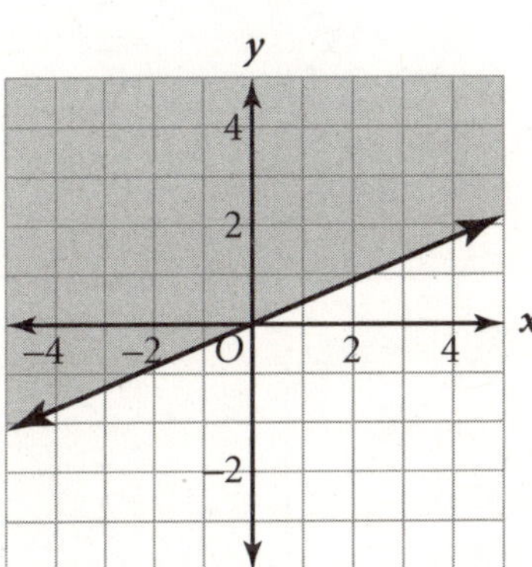

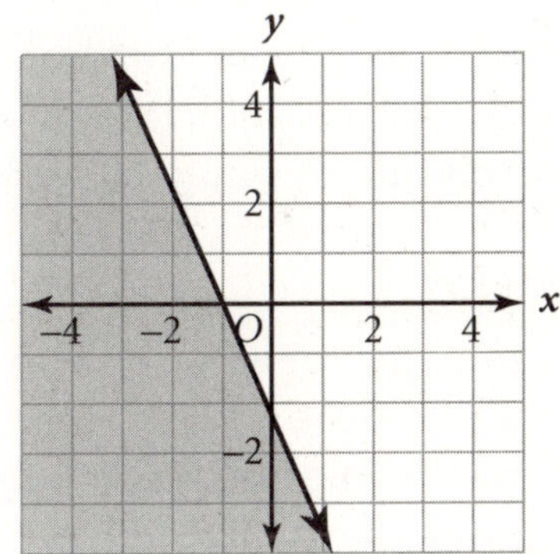

When the two inequalities are graphed on the same coordinate plane, the half planes that are the individual solutions should be put together to give the common solution. The region where the two half planes overlap is the solution to the system. To explain this procedure to students, you can make two transparencies of the two graphs above, and then place one over the other on an overhead projector. A darker shaded region like the one at right will appear. This region is the solution to $\begin{cases} 2x - 5y \le 0 \\ y \le -\frac{1}{4}x - 2 \end{cases}.$

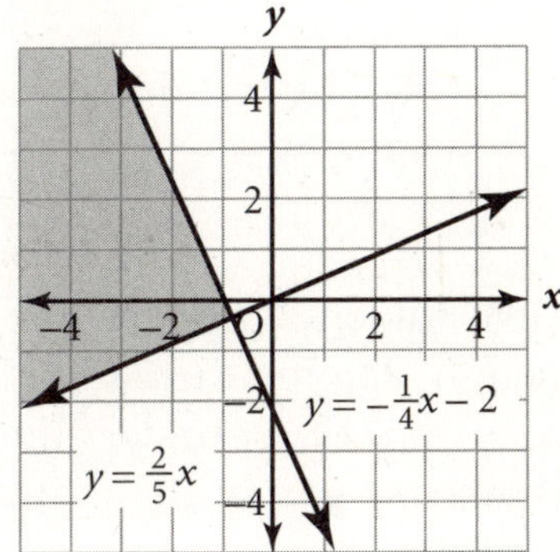

Of all the lessons in a chapter on systems of equations and inequalities, the lesson on systems of inequalities integrates the largest number of skills. The skills needed to display the solution region are all skills that students have learned previously. In fact, the only new skill is the superimposition of two solution regions. Students can use different patterns for shading or different colors to indicate individual solution regions and the common solution region.

In solving application problems involving linear inequalities in two variables, the restrictions $x > 0$ and $y > 0$ are often either stated or implied. It is important not to ignore such restrictions, as they can make the difference between a correct solution region and an incorrect one.

Consider $\begin{cases} y > 0.5x - 2 \\ y < x + 2 \\ x > 0 \\ y > 0 \end{cases}$.

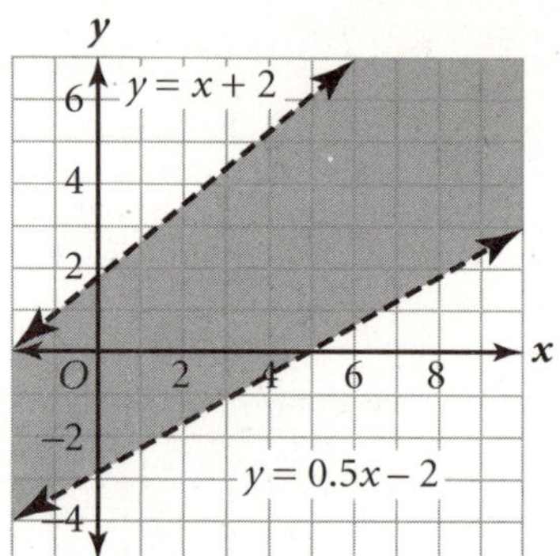

The solution to $\begin{cases} y > 0.5x - 2 \\ y < x + 2 \end{cases}$ is shown at right. Notice, however, that the region extends into all four quadrants of the coordinate plane. If this region is offered as the solution, it would be incorrect.

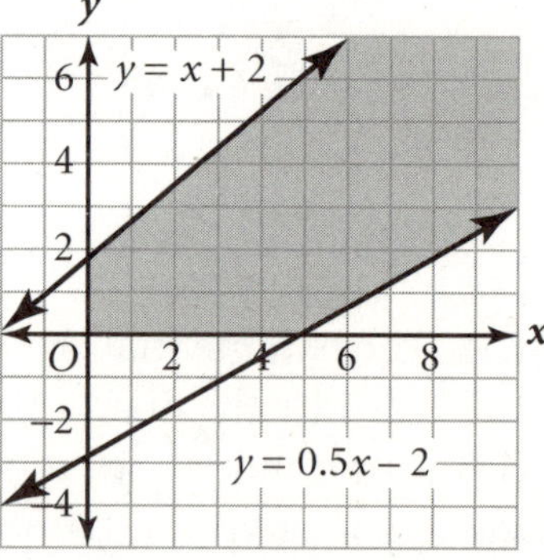

You can make the solution complete by erasing part of it. Simply erase the part that extends into the second, third, and fourth quadrants. The correct solution is shown at right. Any part of the boundary that lies along an axis should then be rendered as a dashed line segment.

Give students enough time to practice the process of sketching a solution region. Remind them to pay attention to the inequality signs involved when they draw the boundary lines, and any explicit or implicit restrictions when they erase part of the solution region.

Consider this problem.

How can I represent the shaded parallelogram at right as a system of linear inequalities in two variables?

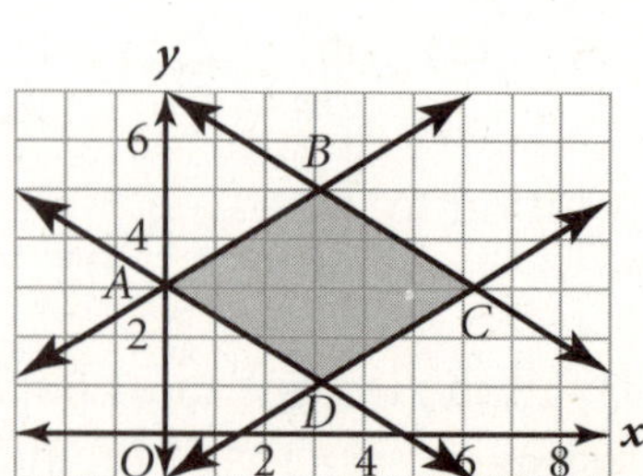

Ask the class to think about a solution strategy.

Encourage all students to participate in making observations. Interactive discussions can help students refine their understanding of concepts. Many observations can be made about the given parallelogram. Although some of them may seem obvious, they are important nonetheless. For example, the observation that the picture is a coordinate plane is obvious but very significant. It allows use of algebra.

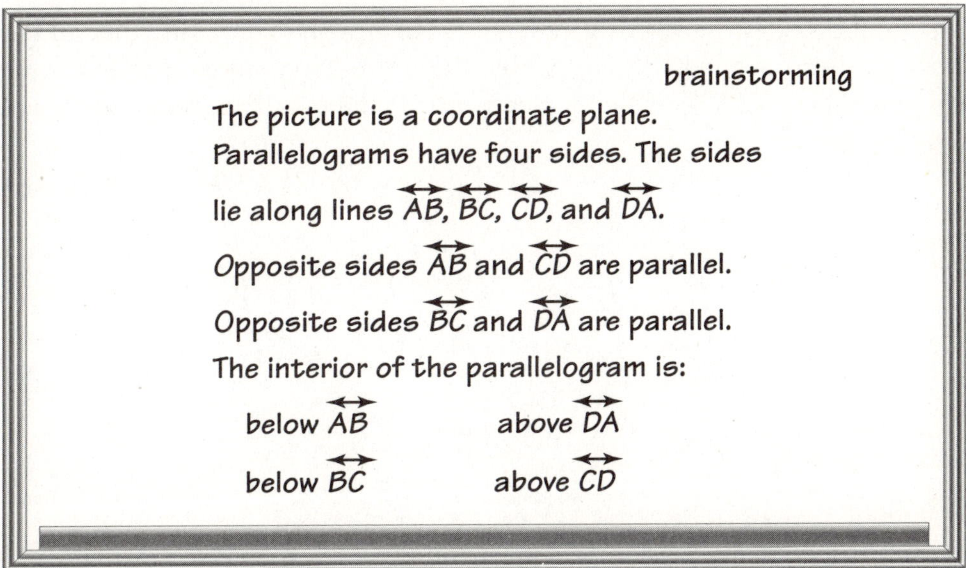

You can move the discussion to a more formal phase, the level in which students devise a plan for using algebra. In the box at right, a strategy that uses equations is devised. Equations are then turned into the needed inequalities. This strategy has the advantage that students can transfer their knowledge about finding an equation for a line.

strategy phase

❶ Find equations for $\overleftrightarrow{AB}$, $\overleftrightarrow{BC}$, $\overleftrightarrow{CD}$, and $\overleftrightarrow{DA}$.

❷ Replace each equality sign ($=$) with an inequality symbol so that the common solution to the system is the region

below $\overleftrightarrow{AB}$ above $\overleftrightarrow{DA}$

below $\overleftrightarrow{BC}$ above $\overleftrightarrow{CD}$

Now students can put their algebra skills to work. If necessary, review Chapter 3, "Finding an Equation for a Line," and reteach this lesson to students. Also have students develop a plan or strategy for gathering the information required. Remind students that they need to use previously learned concepts, such as slope and y-intercept, and the strategy of transforming an equation into an inequality.

$\overleftrightarrow{AB}$	$\overleftrightarrow{CD}$	$\overleftrightarrow{BC}$	$\overleftrightarrow{DA}$
$y = \frac{2}{3}x + 3$	$y = \frac{2}{3}x - 1$	$y = -\frac{2}{3}x + 7$	$y = -\frac{2}{3}x + 3$
below $\overleftrightarrow{AB}$	above $\overleftrightarrow{CD}$	below $\overleftrightarrow{BC}$	above $\overleftrightarrow{DA}$
(use $\leq$)	(use $\geq$)	(use $\leq$)	(use $\geq$)
$y \leq \frac{2}{3}x + 3$	$y \geq \frac{2}{3}x - 1$	$y \leq -\frac{2}{3}x + 7$	$y \geq -\frac{2}{3}x + 3$

The final phase is the verification phase. This involves checking that all the work has been completed successfully. To do this, have students check to see if a point such as $(3, 3)$, which is clearly inside the parallelogram, satisfies all four inequalities.

$$y \leq \frac{2}{3}x + 3 \quad \rightarrow \quad 3 \leq \frac{2}{3}(3) + 3 \quad \rightarrow \quad 3 \leq 5 \ ✔$$

$$y > \frac{2}{3}x - 1 \quad \rightarrow \quad 3 > \frac{2}{3}(3) - 1 \quad \rightarrow \quad 3 > 1 \ ✔$$

$$y \geq -\frac{2}{3}x + 3 \quad \rightarrow \quad 3 \geq -\frac{2}{3}(3) + 3 \quad \rightarrow \quad 3 \geq 1 \ ✔$$

$$y \leq -\frac{2}{3}x + 7 \quad \rightarrow \quad 3 \leq -\frac{2}{3}(3) + 7 \quad \rightarrow \quad 3 \leq 5 \ ✔$$

POLYNOMIALS AND FACTORING

Origin of a Polynomial

Our number system is based on 10. This means that all counting numbers such as 432 can be read and written in terms of powers of 10.

$$\text{four hundred thirty-two}$$

$$4 \times 100 + 3 \times 10 + 2$$

$$\text{or}$$

$$4 \times (10 \times 10) + 3 \times 10 + 2$$

$$\text{or}$$

$$4 \times 10^2 + 3 \times 10^1 + 2 \times 10^0$$

For 432, 4 is in the 10^2 place, 3 is in the 10^1 place, and 2 is in the 10^0 place. The value of the digit 4 in 432 is 4×10^2. Every counting number can be written as a finite sum of products of digits and positive integer powers of 10.

Negative integer powers of 10 represent fractions whose numerators are 1 and whose denominators are positive integer powers of 10. The first three of these are shown below.

$$10^{-1} = \frac{1}{10^1} \qquad 10^{-2} = \frac{1}{10^2} \qquad 10^{-3} = \frac{1}{10^3}$$

Negative powers of 10 are used for decimals that have nonzero digits to the right of the decimal point. The place names for the three powers of 10 given above are stated here.

$$\text{tenths} \qquad \text{hundredths} \qquad \text{thousandths}$$

A decimal that does not represent a counting number can also be written in expanded form. However, the expanded form contains both positive and negative powers of 10. For example, 54.326 is written in expanded form as follows.

$$\underset{5\ \text{tens}}{5 \times 10^1} + \underset{4\ \text{units}}{4 \times 10^0} + \underset{3\ \text{tenths}}{3 \times 10^{-1}} + \underset{2\ \text{hundredths}}{2 \times 10^{-2}} + \underset{6\ \text{thousandths}}{6 \times 10^{-3}}$$

$$54.326 = 5 \times 10^1 + 4 \times 10^0 + 3 \times 10^{-1} + 2 \times 10^{-2} + 6 \times 10^{-3}$$

Notice that the expansion of a nonintegral rational number may involve positive, 0, and negative integer powers of 10.

polynomial: The sum or difference of two or more monomials.

The concept of a **polynomial** is a generalization of the concept of counting number, or integer. In simple terms, you can create a polynomial by writing out a counting number with its powers of 10 and replacing each 10 with a variable such as x.

$$4 \times 10^2 + 3 \times 10^1 + 2 \times 10^0 \qquad \textit{counting number}$$
$$4 \times x^2 + 3 \times x^1 + 2 \times x^0 \qquad \textit{polynomial}$$
$$4x^2 + 3x^1 + 2x^0 \qquad \textit{or simply } 4x^2 + 3x + 2$$

In the case of a counting number, only the *digits*, counting numbers from 0 to 9 inclusive, are used as multipliers of powers of 10. In the case of a polynomial, any real number can be a multiplier of the variable. Each of the expressions below is a polynomial in x. These polynomials are, in fact, equivalent. They differ only in that the terms of the second polynomial have been made more explicit by writing plus signs ($+$) between any two terms.

$$-x^3 + 5x^2 - 4x + 7$$
$$(-1)x^3 + 5x^2 + (-4)x + 7$$

The transition from counting numbers to polynomials illustrates the fact that certain mathematical concepts share common meanings. Exploring these conceptual relationships can help students enhance their mathematical understanding. After all, this type of exploratory work is what mathematicians do. Mathematicians often create abstract objects involving variables by looking at familiar objects such as numbers. When the new objects are created, mathematicians ask questions such as the following:

Mathematical Inquiries

- How can we add, subtract, multiply, and divide such new objects?

- If a polynomial in one variable is set equal to 0, then it becomes an equation. What can be discovered by studying solution methods for such equations?

- What sorts of real-world problems can be solved by using these new objects?

- What conjectures about polynomials can be discovered and how can such conjectures be proved?

Students can also ask similar questions when they study mathematics topics such as polynomials. The range of possible explorations and inquiries depends on how much a student knows about a topic and how much more the students wants to know.

Since one of the goals of mathematics education is to engage students in mathematical discourse, we must first define the mathematical terms that we expect our students to use in their discussion. Ambiguity is a serious potential source of misunderstanding. To minimize ambiguity, it is necessary to define mathematical terms precisely. In our previous discussion, we developed the concept of a polynomial from the concept of a counting number. To talk about polynomials, we now need to agree on the vocabulary that will be used in our discussion of polynomials.

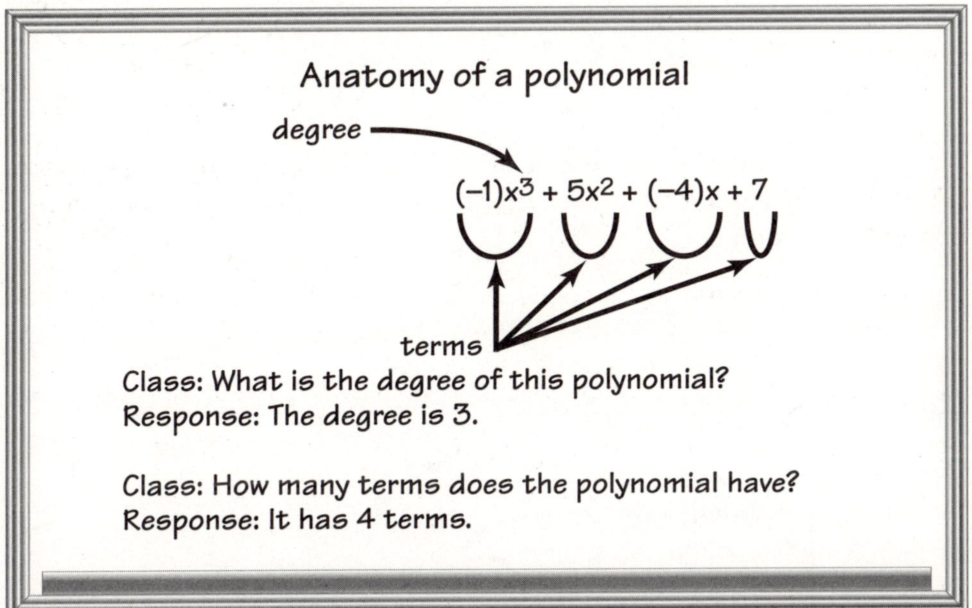

A polynomial is a sum of terms. Each of these terms is also called a **monomial**, a product of a real number and a nonnegative power of the variable.

You can classify a polynomial by its **degree**.

monomial: An algebraic expression that is either a constant, a variable, or a product of a constant and one or more variables.

degree of a polynomial: The degree of a polynomial in one variable is determined by the exponent with the greatest value within the polynomial.

Classification of a Polynomial by Degree

linear: degree 1   quadratic: degree 2   cubic: degree 3

quartic: degree 4   quintic: degree 5

term: A number, a variable, or a product or quotient of numbers and variables that is added or subtracted in an algebraic expression.

You can also classify a polynomial by its number of **terms.** Polynomials with more than three terms are not given special names.

Classification of a Polynomial by Number of Terms

monomial: 1 term   binomial: 2 terms   trinomial: 3 terms

Notice that classifying a polynomial by degree 2 or 3, for example, is somewhat like classifying a counting number as in the hundreds or in the thousands. Of course, this rough comparison cannot be taken too far.

Mathematical writing should be simplified whenever possible. For example, a variable that is raised to the first power does not need to be written with the exponent of 1, and a variable that is raised to the 0 power can be omitted altogether.

$$3x^1 = 3x \quad \text{and} \quad 2x^0 = 2 \text{ since } x^0 = 1$$

coefficient: A number that is multiplied by a variable.

Thus, the polynomial $4 \times x^2 + 3 \times x^1 + 2 \times x^0$ is written as $4x^2 + 3x + 2$. If a polynomial has 0 as a **coefficient** of a term, that term need not be written in the sum.

Vocabulary terms that will be needed can be illustrated on the whiteboard as follows.

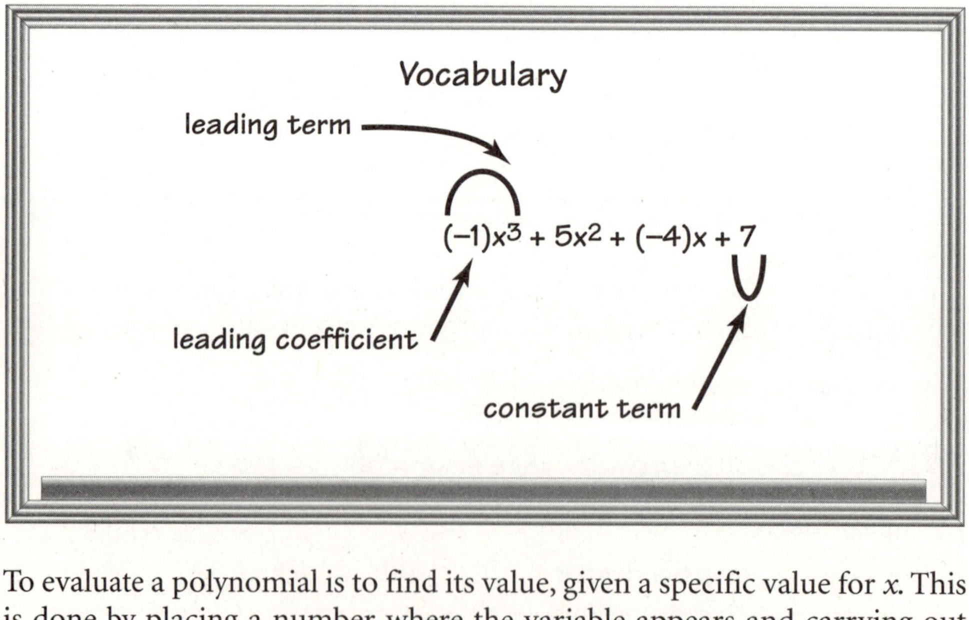

To evaluate a polynomial is to find its value, given a specific value for *x*. This is done by placing a number where the variable appears and carrying out the arithmetic according to the order of operations.

> ### *Order of Operations*
>
> ❶ Perform multiplication with exponents.
>
> ❷ Multiply and divide from left to right.
>
> ❸ Add or subtract from left to right.

Consider $3x^4 - 2x^3 + 5x^2 + 7x + 3$ with x replaced by 3.

$$
\begin{aligned}
❶\quad 3(3)^4 - 2(3)^3 + 5(3)^2 + 7(3) + 3 &= 3(81) - 2(27) + 5(9) + 7(3) + 3 \\
❷\quad &= 243 - 54 + 45 + 21 + 3 \\
❸\quad &= 189 + 45 + 21 + 3 \\
&= 258
\end{aligned}
$$

Addition and Subtraction

When students learn to find a sum, such as $432 + 75$, they are taught to line up decimal places. The short form that students learn to use is shown at left below. The form using powers of 10 is shown at right below. Notice in the sum $432 + 75$ that $7 + 3 = 10$. The 0 is written in the 10^1 place and 1 is brought over to the 10^2 place.

$$
\begin{array}{r}
{}^{1}4\ 3\ 2 \\
+\ \ \ 7\ 5 \\
\hline
5\ 0\ 7
\end{array}
\quad\longrightarrow\quad
\begin{array}{r}
4 \times 10^2 + 3 \times 10^1 + 2 \times 10^0 \\
+\qquad\qquad 7 \times 10^1 + 5 \times 10^0 \\
\hline
(1 + 4) \times 10^2 + 0 \times 10^1 + 7 \times 10^0
\end{array}
$$

The addition with powers of ten should give students a cue as to how to add two polynomials.

> ### Addition of Two Polynomials in the Same Variable
>
> To find the sum of two polynomials in the same variable, arrange the terms so that corresponding powers of the variable are grouped, add corresponding coefficients, then write the result.

The sum of $4x^2 + 3x + 2$ and $7x + 5$ is shown in vertical format below.

$$
\begin{array}{r}
4x^2 + 3x + 2 \\
+\qquad\quad 7x + 5 \\
\hline
4x^2 + 10x + 7
\end{array}
$$

This is a very practical and convenient way to arrange the addition of polynomials. Notice that constant terms are aligned, x terms are aligned, and x^2 terms are aligned.

Notice the similarity between this addition and the addition of 432 and 75. This form of addition of polynomials is recommended for beginning algebra students because of its similarity with the format used to add counting numbers.

An alternative format to add two polynomials is the horizontal format. The sum of $4x^2 + 3x + 2$ and $7x + 5$ in horizontal format is shown below.

$$(4x^2 + 3x + 2) + (7x + 5) = 4x^2 + 3x + 2 + 7x + 5 \quad \leftarrow \text{Commutative Property of Addition}$$

$$= 4x^2 + 3x + 7x + 2 + 5 \quad \leftarrow \text{Commutative Property of Addition}$$

$$= 4x^2 + (3 + 7)x + 2 + 5 \quad \leftarrow \text{Distributive Property}$$

$$= 4x^2 + 10x + 7 \quad \leftarrow \text{Add numbers}$$

One purpose of using this format is to help students see how algebraic properties are used to find the sum. You might view this format as formal and perhaps difficult. However, this format shows that mathematical work is based on organized and legitimate mathematical concepts. Students should learn to perform addition in both formats so that they can appreciate the value and purpose of each format.

With practice, students can perform the addition above mentally, and simply write down the sum.

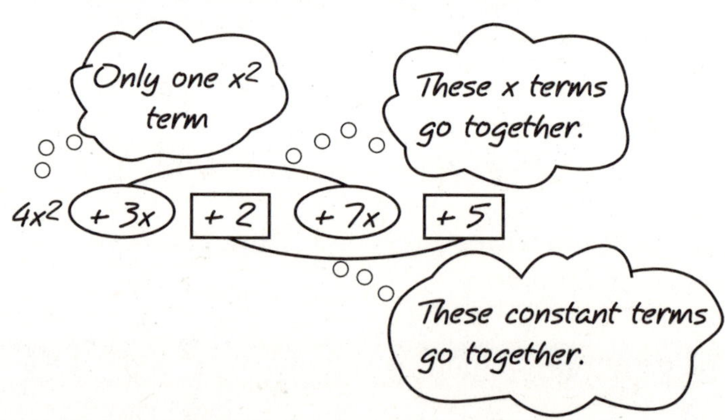

The diagram at right suggests the mental processes involved. The student makes observations about the terms in the sum. Notice that there are three observations and three mental steps in finding the sum $4x^2 + 3x + 2 + 7x + 5$. As before, that sum is $4x^2 + 10x + 7$.

We have emphasized the fact that mathematics skills are highly interrelated. For example, to perform operations on polynomials, students need to know how to add rational numbers, especially integers. For example, to find the

$$
\begin{array}{r}
-8x + \qquad -7 \\
+ \quad 3x + \qquad 12 \\
\hline
(-8 + 3)x + (-7 + 12) \\
-5x + \qquad 5
\end{array}
$$

sum of $-8x - 7$ and $3x + 12$, students need to know how to find the sums $-8 + 3$ and $-7 + 12$.

The following is an important property about the addition of polynomials. Ask students to rephrase this property in their own words to make sure that they understand its meaning.

Closure Property of Addition

The sum of two polynomials in the same variable is always another polynomial in that variable. That is, addition of polynomials is *closed*.

Students can solve a problem like the one below by using simple polynomial addition and some geometry.

What is the total distance around this figure? The three rectangles have the same dimensions and the two squares have the same dimensions.

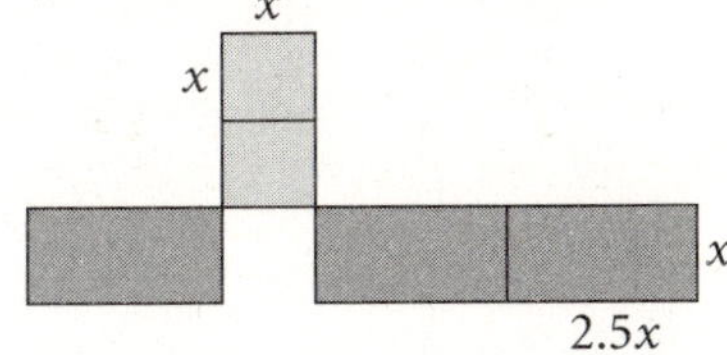

Direct observation shows that there are eight sides in the two squares. Each has length x. But two of the sides coincide and do not count as distance around the entire figure. Therefore, the distance around the two squares equals 6 x's. There are 3 rectangles each having a width of x. But two of the widths coincide and do not count. You will need 4 more x's. There are 6 lengths, each $2.5x$ units long. Since no lengths coincide, you will need all of them.

$$\text{total distance} = 6x + 4x + 6(2.5x) = 25x$$

If the cue for how to add two polynomials comes from addition of numbers, then perhaps the cue for how to subtract one polynomial from another will also come from subtraction of numbers.

Consider this problem.

Suppose that you have $100 in the bank. You withdraw $20 and then withdraw $30 from the account. How much do you still have in the bank?

$$
\begin{array}{cc}
\text{sequential withdrawals} & \text{combined withdrawals} \\
100 - 20 - 30 = 80 - 30 & 100 - (20 + 30) = 100 - 50 \\
= 50 & = 50
\end{array}
$$

Using either calculation, you will have $50 still in the account.

From this banking situation, you can conclude that the sequential withdrawals, $20 and $30, are numerically the same as the combined withdrawals, the sum of $20 and $30.

$$-20 - 30 = -(20 + 30), \text{ and } -(20 + 30) = -20 - 30$$

You might be wondering what this banking situation has to do with subtracting polynomials. Here is the explanation.

To subtract one polynomial from another polynomial, add the opposite of the second polynomial to the first polynomial. To find the opposite of a polynomial, change the sign of each term in it from $(+)$ to $(-)$ or $(-)$ to $(+)$.

$$\begin{array}{cc} \text{polynomial} & \text{opposite} \\ 13x^2 - 5 & -(13x^2 - 5) = -13x^2 + 5 \end{array}$$

Consider $(2x^2 - x + 5) - (3x^2 - 5x - 3)$.

Vertical format for the subtraction is shown below. Notice that when the subtraction problem is written as an addition problem, each sign in $3x^2 - 5x - 3$ is changed from $(+)$ to $(-)$ or $(-)$ to $(+)$.

$$\begin{array}{ccc}
\text{subtraction} & & \text{addition} \\
2x^2 - x + 5 & & +2x^2 \;-\; x \;+ 5 \\
\underline{-\ (3x^2 - 5x - 3)} & \longrightarrow & \underline{+\ -3x^2 \;+ 5x \;+ 3} \\
& & -x^2 \;+ 4x \;+ 8
\end{array}$$

The difference equals $-x^2 + 4x + 8$.

Multiplication

When students learn how to multiply two counting numbers, they learn to carry out procedures. Some students, however, do not pay much attention to how or why these procedures work. The multiplication problem shown at right shows how students carry out procedures. Notice that the student performs two multiplications and one addition.

$$\begin{array}{r}
23 \\
\times\ 17 \\
\hline
161 \leftarrow 7 \times 23 \\
+\ 230 \leftarrow 10 \times 23 \\
\hline
391 \leftarrow 161 + 230
\end{array}$$

The same multiplication can be done by changing each counting number into a power of ten, as shown below.

$$\begin{aligned}
23 \times 17 &= (2 \times 10^1 + 3)(1 \times 10^1 + 7) \\
&= (2 \times 10^1 + 3)(7) \qquad + \qquad (2 \times 10^1 + 3)(1 \times 10^1)
\end{aligned}$$

first row: 161   *second row: 230*

When these products are expanded, the products shown above are obtained. The final answer is the sum of 161 and 230.

Notice that the product $(2 \times 10^1)(1 \times 10^1)$ involves multiplying two powers of 10. This product requires that the student write 23 in the second row shifted to the left one decimal place. That is, the student writes 23 in the second row but not directly under 161 in the first row.

How does all of this relate to multiplying one polynomial such as $3x - 2$ by another polynomial such as $4x + 5$? The answer is that multiplication of polynomials involves skills from multiplication of counting numbers. We recommend that the students practice a number of problems like the ones given in each stage below before attempting to multiply $(3x - 2)(4x + 5)$.

Stage 1 Learn how to find small products, such as $(3x)(4x)$.

$$(3x)(4x) = 3 \cdot 4 \cdot x \cdot x = 12x^2 \leftarrow x \cdot x = x^2$$

Stage 2 Learn how to multiply a polynomial by a monomial, such as $3x(4x + 5)$.

$$3x(4x + 5) = 3x(4x + 5) = (3x)(4x) + (3x)(5) = 12x^2 + 15x$$

These come from Stage 1.

Stage 3 Learn how to put together what was learned in Stages 1 and 2.

These come from Stage 2.

$$(3x - 2)(4x + 5) = 3x(4x + 5) - 2(4x + 5)$$
$$= 12x^2 + 15x - 8x - 10$$
$$= 12x^2 + 7x - 10$$

You can show students a compact form of multiplication of polynomials that is very similar to the form they would use when multiplying counting numbers.

$$
\begin{array}{r}
3x - 2 \\
\times\ \ 4x + 5 \\
\hline
15x - 10 \quad \longleftarrow\ 5(3x - 2) \\
+\ 12x^2 - 8x \qquad\ \longleftarrow\ 4x(3x - 2) \\
\hline
12x^2 + 7x - 10
\end{array}
$$

The first row comes from $5(3x - 2)$ and the second row comes from $4x(3x - 2)$. Notice also that when the second row is written, it is shifted to the left. This shift keeps corresponding powers of x aligned in the same way that a shift to the left keeps corresponding powers of 10 aligned in multiplication of numbers.

The following schematic representation shows the same multiplication problem as a series of connected steps. Some teachers may want to reproduce this schematic representation one step at a time on the whiteboard. Alternatively, some teachers may make a transparency and uncover parts of it as they guide a discussion.

$$(3x - 3)(4x + 5)$$

Polynomial times monomial $(3x - 2)(4x) + (3x - 2)(5)$

Monomial times monomial $(3x)(4x) + (-2)(4x) + (3x)(5) + (-2)(5)$

$$12x^2 + (-8x) + 15x + (-10)$$

$$12x^2 + 7x - 10$$

Just as the product of two integers is another integer, the product of two polynomials is another polynomial.

> ### Multiplication of Polynomials
>
> If you multiply one polynomial in one variable by another polynomial in that variable, you will always get another polynomial in that variable. That is, multiplication of polynomials is closed.

The discussion of polynomial multiplication is quite extensive. At the least, you can see that mathematicians use numerical skills as a starting point for defining algebraic procedures and skills. In short, addition, subtraction, and multiplication of polynomials are naturally defined by looking at how addition, subtraction, and multiplication of numbers work.

A discussion of multiplication might conclude with a simple geometry problem. The figure below contains the data for this problem.

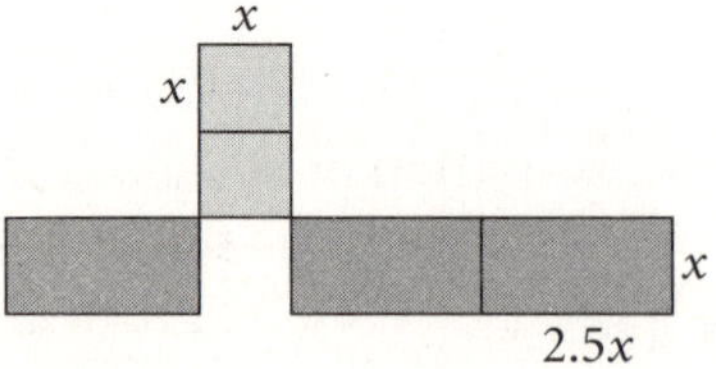

Find the total area of this figure. The two squares have the same dimensions and the three rectangles have the same dimensions.

Observe that the figure consists of two squares, each x units on one side. The area of one square is x^2. You will need two of these. The figure also contains three rectangles each x units by $2.5x$ units. The area of one of these is $2.5x^2$. You will need three of these.

$$\text{total area} = 2x^2 + 3(2.5x^2) = 9.5x^2$$

Division

Students should learn that not every counting number is divisible by every other counting number. For example, when one pizza is cut into 6 slices, no counting number represents the part of the entire pizza that each slice represents. Division of one counting number by another can be explained by the concept of fraction.

Consider $\frac{38}{5}$.

$$\frac{38}{5} = \frac{5 \cdot 7 + 3}{5} = \frac{5 \cdot 7}{5} + \frac{3}{5} = 7 + \frac{3}{5} = 7\frac{3}{5}$$

The quotient $\frac{38}{5}$ is represented by a mixed number, $7\frac{3}{5}$, the sum of a whole number and a fraction. A mixed number is not a counting number.

Operations with polynomials also include division. To teach division of polynomials, we suggest to follow these stages.

Stage 1 Learn to divide one monomial by another monomial.

Stage 2 Learn to divide one polynomial by a monomial.

When students progress through Stage 1 in the multiplication process, they learn to use what is called the Product Law of Exponents.

Product Law of Exponents

If $a \neq 0$ and m and n are counting numbers, then $a^m \cdot a^n = a^{m+n}$.

In words, when they multiply two powers with the same base, they add the exponent of one factor to that of the other factor.

When students progress through Stage 1 in the division process, they will learn a new law of exponents. It is the Quotient Law of Exponents.

Quotient Law of Exponents

If $a \neq 0$ and m and n are counting numbers, then $\frac{a^m}{a^n} = a^{m-n}$.

In words, when they divide two powers with the same base, they subtract the exponent of the denominator from that of the numerator.

Consider $\dfrac{15x^5}{12x^3}$, a quotient of monomials.

Think of $15x^5$ as $15 \cdot x^5$ and think of $12x^3$ as $12 \cdot x^3$. Now the problem looks like the result of multiplying two fractional expressions as shown below. If you work on the two fractional expressions separately, you can use division of numbers and the Quotient Law of Exponents to find each quotient.

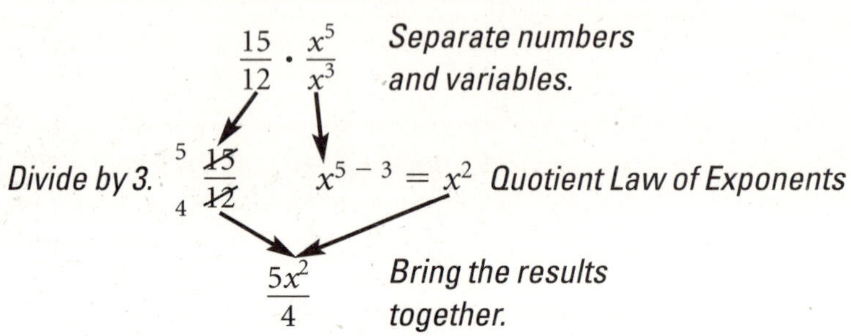

Notice in the diagram above, that students turn one problem into a pair of smaller problems. This is a good strategy for students who are beginning to learn how to divide. Also, separating the numerical part of the problem from the variable part of it may help students simplify the work with division of monomials.

Once students have learned how to perform division of two monomials, they can transfer this knowledge to division of a polynomial by a monomial.

Consider $\dfrac{15x^5 - 6x^4}{12x^3}$, a quotient of a binomial by a monomial.

This quotient looks like the difference of two fractional expressions with the same denominator.

$$\dfrac{15x^5 - 6x^4}{12x^3} = \dfrac{15x^5}{12x^3} - \dfrac{6x^4}{12x^3} \qquad \textit{Make two smaller problems}$$

$$\dfrac{5x^2}{4} - \dfrac{x}{2} \qquad \textit{Use the skills from Stage 1.}$$

You can also find the quotient by using factoring. To do this, recall what is meant by the **greatest common factor (GCF)** of two numbers and of two monomials. The greatest common factor of $15x^5$ and $6x^4$ is the greatest common factor of 15 and 6 times the highest power of x that is common to x^5 and x^4. The GCF is $3x^4$.

Consider $\dfrac{15x^5 - 6x^4}{12x^3}$ again.

The following whiteboard demonstration shows the simplification of this problem. The first step is the numerical part of the work. Divide numerator and denominator by the greatest common factor of the coefficients, that is, divide each coefficient by 3.

$$\frac{15x^5 - 6x^4}{12x^3}$$

Class: Would someone come to the whiteboard and divide from the numerator and denominator the greatest common numerical factor?

Response: $\dfrac{\overset{5}{\cancel{15}}x^5 - \overset{2}{\cancel{6}}x^4}{\underset{4}{\cancel{12}}x^3} = \dfrac{5x^5 - 2x^4}{4x^3}$

The second part of the simplification involves dividing all terms by the highest power of x they all have. That power is x^3.

Class: Would someone come to the whiteboard and divide from the numerator and denominator the highest power of x possible?

Response: $\dfrac{5x^{\cancel{5}\,2} - 2x^{\cancel{4}\,1}}{4x^{\cancel{3}}} = \dfrac{5x^2 - 2x}{4}$

Integer Factorization

Multiplication of two numbers is the process of writing a single number as the product of two numbers.

pair of numbers in a product   single number in the product

3×4 $=$ 12

Factoring a number is the reverse process. A single number is written as a multiplication of two numbers.

single number   pair of numbers in a product

12 $=$ 3×4

From the diagram at right, you can see that factoring and division are related processes. You can also see from the diagram that more than one grouping is possible. For example, you could also separate the group of 12 pencils into 2 groups of 6 pencils each.

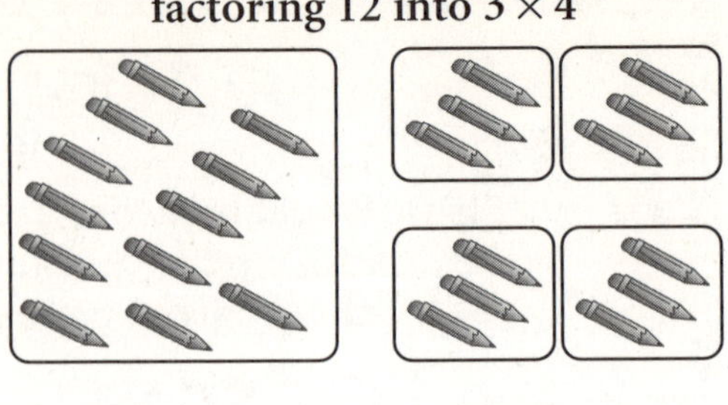

Some integers can be factored in more than one way. Such integers are called *composite numbers*.

$$24 \rightarrow 1 \times 24, 2 \times 12, 3 \times 8, 4 \times 6$$

Some integers can only be factored in one way. Such numbers are called *prime numbers*.

$$5 \rightarrow 1 \times 5 \text{ or } 5 \times 1 \quad \text{and} \quad 31 \rightarrow 1 \times 31 \text{ or } 31 \times 1$$

The following example illustrates how to write a *prime factorization* of a counting number. The prime factorization of a number is a product consisting of powers of prime numbers only. To begin the process, use multiplication facts. Since $6 \times 20 = 120$, start the process by writing 120 as 6×20.

Consider 120.

Write 120 as 6×20.

Write 6 and 20 as products.

Write 4 as a product.

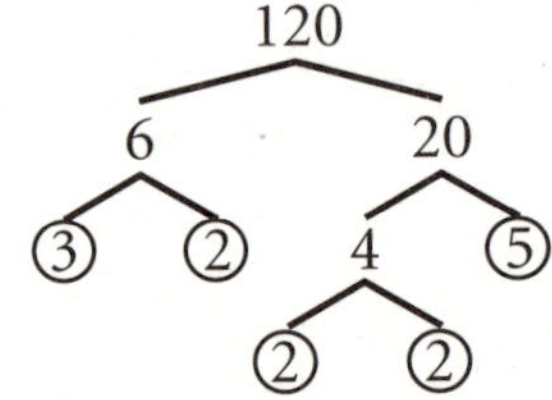

The diagram above is called a *factor tree*. Notice that a branch of the tree ends when a prime number is found as a factor. The prime factorization of 120 is $2^3 \times 3^1 \times 5^1$.

You can also begin the process by writing 4×30 instead of 6×20. The factor tree below illustrates this case.

Write 120 as 4×30.

Write 4 and 30 as products.

Write 6 as a product.

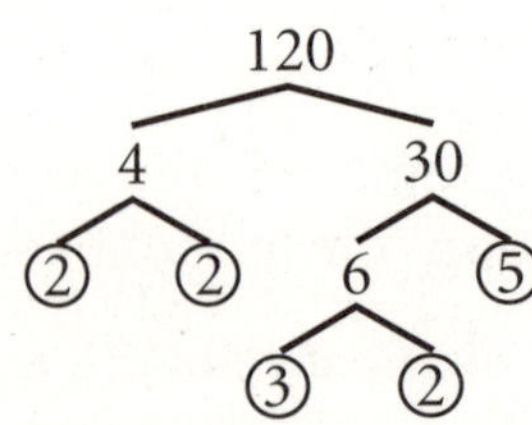

Notice that the prime numbers at the bottom of the factor tree are exactly the same as before. Again, $120 = 2^3 \times 3^1 \times 5^1$.

The prime factorization of a number is unique except for the order in which the individual factors are written.

Simple Polynomial Factorization

The notion of factoring can be transferred from the world of numbers to the world of polynomials.

product of two polynomials		single polynomial
$(3x - 2)(4x + 5)$	$=$	$12x^2 + 7x - 10$

single polynomial		product of two polynomials
$12x^2 + 7x - 10$	$=$	$(3x - 2)(4x + 5)$

Unfortunately, the straightforward multiplication facts that apply to numbers (such as $4 \times 30 = 120$) cannot be used to split a polynomial into a pair of polynomials that make the required product.

Learning how to factor polynomials is an essential algebraic skill that has multiple uses in application problems. Factoring polynomials also provides the opportunity to learn other problem-solving skills.

Factoring methods usually progress from simpler to more complex polynomials. One of the first factoring methods is the application of the Distributive Property.

Distributive Property

For all numbers a, b, and c, $ab + ac = a(b + c)$.

In words, when a is a factor common to both ab and ac, you may factor out a. Then you may write the product of a and what is left of ab and bc after a is removed.

Consider $2x + 4$.   *2 is a common factor.*

$$2x + 4 = \boxed{2} \cdot x + \boxed{2} \cdot 2$$

You can write $2x + 4$ as $2(x + 2)$.

Consider $2x^2 + 4x$.   *2x is a common factor.*

$$2x^2 + 4x = \boxed{2x} \cdot x + \boxed{2x} \cdot 2$$

The common factor, $2x$, will be one of the expressions in the factorization. What remains of each expression in the sum will be the other factor.

$2x$: factor	$x + 2$: factor

Factorization: $2x(x + 2)$

With sufficient practice, students can devise a procedure to factor many polynomials. You can help students do this by modeling a whiteboard demonstration with guiding questions.

In the following example, students find that $3x^2$ is the greatest common factor of all the terms.

Consider $12x^4 + 18x^3 + 21x^2$.

$$12x^4 + 18x^3 + 21x^2$$

Class: Look at the numerical parts of the terms. What is the greatest number that is a common factor for all of them?

Response: The greatest number that will divide 12, 18, and 21 is 3.

Class: Look at the variable parts of the terms. What is the highest power of x common to all of them?

Response: The highest power of x common to all terms is 2.

Class: Write $3x^2$. Then write a pair of parentheses.

$$3x^2 (\qquad\qquad)$$

Class: What is left of the terms after $3x^2$ is taken out?

$$3x^2 \cdot \boxed{4x^2} \qquad 3x^2 \cdot \boxed{6x^1} \qquad 3x^2 \cdot \boxed{7}$$

Class: Write the circled expressions inside the parentheses.

$$3x^2 (4x^2 \quad + \quad 6x \quad + \quad 7)$$

Therefore, you can write $12x^4 + 18x^3 + 21x^2$ as $3x^2(4x^2 + 6x + 7)$.

If students are keeping a journal that contains class notes, this discussion is one they may want to include in their journals. In particular, students might want to jot down the guidelines and questions that help them find the final result.

Students should be able to use the factorization process illustrated above, given sufficient guidance and practice.

More Complicated Factoring

Suppose that you are given $x^2 + 5x + 6$. Consider the following reasoning.

$$x^2 + 5x + 6 = x^2 + \overbrace{2x + 3x}^{\text{another name for }5x} + 6$$

Factor $x^2 + 2x$

Factor $3x + 6$

$$= x(x + 2) + 3(x + 2)$$

$$= (x + 3)(x + 2)$$

Observe that $x + 2$ is common

This shows that $x^2 + 5x + 6$ can be factored as $(x + 3)(x + 2)$.

Notice that the middle term, *5x*, was rewritten and then the resulting polynomial was factored by grouping terms and then by factoring the greatest common factor in each group.

Consider $x^2 + 11x + 18$.

Shown below are several ways to rewrite $x^2 + 11x + 18$.

$$x^2 + 5x + 6x + 18 \quad x^2 + 4x + 7x + 18 \quad x^2 + 10x + 1x + 18$$
$$x^2 + 9x + 2x + 18$$

Which of these, if any, can be used to factor $x^2 + 11x + 18$? The selection of a candidate demands that the two coefficients of *x* have 11 as the sum. All the candidates satisfy this requirement. However, the product of those numbers must be 18. Only one of the candidates above satisfies this requirement.

$$x^2 + 9x + 2x + 18 = x(x + 9) + 2(x + 9) = (x + 2)(x + 9)$$

Therefore, $x^2 + 11x + 18$ can be factored as $(x + 2)(x + 9)$.

If you are curious about confirming that none of the other candidates will work out, here is part of what you might do.

$$x^2 + 5x + 6x + 18 = x(x + 5) + 6(x + 3) \quad \leftarrow \text{ no common factor}$$

Similarly, testing the other possibilities will give two groups of terms with no common factor.

We now review the factorization process.

- Both $x^2 + 5x + 6$ and $x^2 + 11x + 18$ have the same general form

$$x^2 + bx + c, \text{ where } b \text{ and } c \text{ are numbers}$$

- To factor an expression of the form $x^2 + bx + c$, we should look for a pair of factors of *c* whose sum is *b*.

In the case of $x^2 + 5x + 6$, the factors 2 and 3 of 6 have a sum of 5. In the case of $x^2 + 11x + 18$, the factors 2 and 9 of 18 have a sum of 11.

To involve students in solving factorization problems, consider various approaches.

- Show students a problem in the form of a challenge or puzzle, such as the one shown in the following whiteboard illustration.

- Show students the problem as a geometry problem.

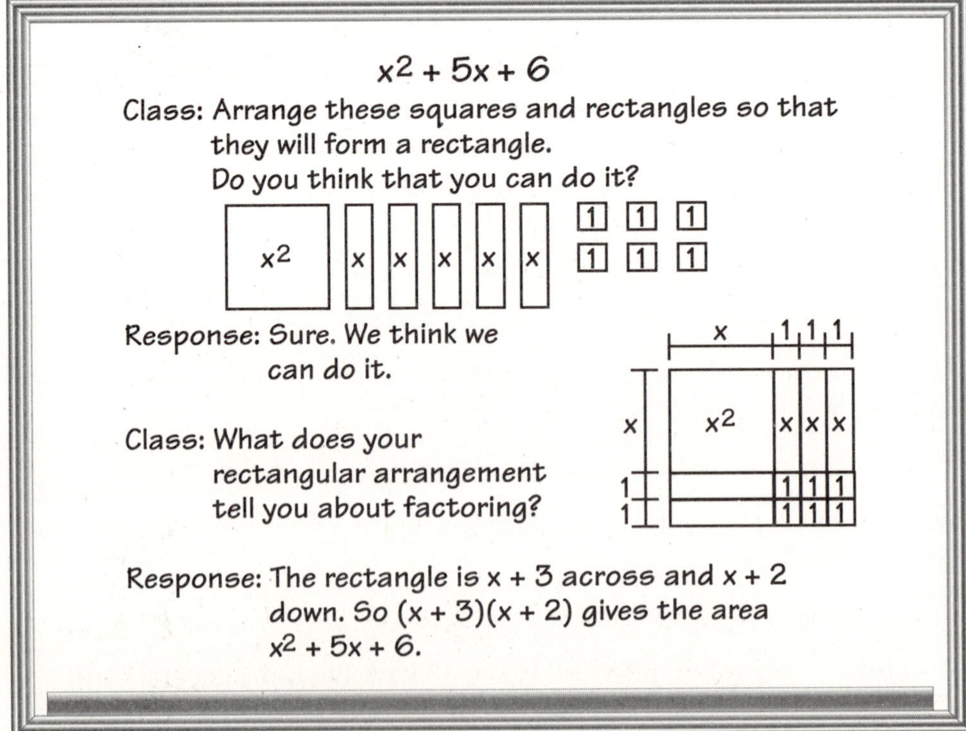

In order to use this approach, you need squares to represent x^2, rectangles to represent x, and small squares to represent 1. These figures need to have proportions that allow for making the necessary arrangements. Many schools have the resources that support this kind of activity.

- Finally, you can show students the problem and ask them to use trial-and-error and the *process of elimination.*

Factors of 6	Sum	Product	Algebra product	Multiplication
1 and 6	$1 + 6 = 7$	6	$(x + 1)(x + 6)$	$x^2 + 7x + 6$
2 and 3	$2 + 3 = 5$	6	$(x + 2)(x + 3)$	$x^2 + 5x + 6$
3 and 2	$3 + 2 = 5$	6	$(x + 3)(x + 2)$	$x^2 + 5x + 6$
6 and 1	$6 + 1 = 7$	6	$(x + 6)(x + 1)$	$x^2 + 7x + 6$

The first column of the table contains all possible positive integer factors of 6. Therefore, if $x^2 + 5x + 6$ can be factored using integers, then one of the rows in the table must give the correct factorization. Using the process of elimination, students will be able to rule out the candidates in the first and fourth rows of the table. Rows 2 and 3 are really the same. They give $x^2 + 5x + 6$ as the product. Therefore, $x^2 + 5x + 6 = (x + 2)(x + 3)$ or $(x + 3)(x + 2)$.

We are not recommending to use all of the approaches presented here. However, if some students find a particular approach difficult to understand, you should be able to explain a different approach. The goal is to maximize students' understanding, not the number of teaching approaches.

Continuing the Study of Factorization

Many textbooks introduce factoring an expression of the form $ax^2 + bx + c$ by presenting different cases. The following four cases illustrate how factoring evolves from simplest to more complex examples.

$$ax^2 + bx + c$$

Case 1 $a = 1$, b is positive, and c is positive.   **Example:** $x^2 + 5x + 6$

Case 2 $a = 1$, b is negative, and c is positive.   **Example:** $x^2 - 5x + 6$

Case 3 $a = 1$, b is negative, and c is negative.   **Example:** $x^2 - 5x - 6$

Case 4   The coefficients can be anything.   **Example:** $10x^2 + 19x + 6$

Enough has been said about Case 1. Notice in Cases 2 and 3 that both positive and negative numbers come into play. When students begin to study cases such as Case 3 and Case 4, they should have developed a good understanding of the use of trial-and-error and testing sums and products of factors of numbers.

Consider $x^2 - 5x + 6$.

If you look at this expression in the same way that you looked at $x^2 + 5x + 6$ earlier, you can transfer skills previously learned and adapt these skills to the new case.

$$x^2 - 5x + 6 = x^2 - 2x - 3x + 6$$
$$= x(x - 2) - 3(x - 2)$$
$$= (x - 3)(x - 2)$$

In other words, to factor $x^2 - 5x + 6$, look for two negative numbers that are factors of 6 and have -5 as their sum.

Consider $x^2 - 5x - 6$.

If you look at this expression in the same way that you looked at $x^2 - 5x + 6$ earlier, you will see that additional modifications to skills previously learned will lead you to the correct factorization.

$$x^2 - 5x - 6 = x^2 + x - 6x - 6$$
$$= x(x + 1) - 6(x + 1)$$
$$= (x + 1)(x - 6)$$

In this case, you need a positive number and a negative number, 1 and -6, in order to get -6 as a product and -5 as a sum.

Based on the cases discussed so far, you can help students make the following generalization.

To factor an expression of the form $x^2 + bx + c$, find two numbers, positive or negative, that have c as product and b as sum.

Finishing the Study
of Factorization

In the previous section, students learned how to factor expressions of the form $ax^2 + bx + c$, regardless of whether a is 1 or some other number.

Consider $10x^2 + 19x + 6$.

Make a table that contains factors of 10 and factors of 6. In this table, there are many combinations of numbers that need to be tested. Which combination of numbers will work? The first row of the table gives the product below.

Factors of 10		Factors of 6	
p	q	r	s
1	10	1	6
1	10	6	1
1	10	2	3
1	10	3	2
10	1	1	6
10	1	6	1
10	1	2	3
10	1	3	2

$$1 \text{ and } 10$$
$$\downarrow \qquad \downarrow$$
$$(1x + 1)(10x + 6)$$
$$\uparrow \qquad \uparrow$$
$$1 \text{ and } 6$$

$$(1x + 1)(10x + 6) = 10x^2 + 16x + 6$$

Because $10x^2 + 16x + 6 \neq 10x^2 + 19x + 6$, row 1 of the table above does not deliver the correct combination of numbers for the factorization. There are seven more rows in the table. You could write seven more products, multiply, and then compare the resulting polynomial with $10x^2 + 19x + 6$. In fact, none of the rows in the table will give the correct combination.

However, the number 10 can be factored by using 2 and 5. The table at right shows all the combinations when 2 and 5 are used instead of 1 and 10. The fourth row of the table gives the product below.

Factors of 10		Factors of 6	
2	5	1	6
2	5	6	1
2	5	2	3
2	5	3	2
5	2	1	6
5	2	6	1
5	2	2	3
5	2	3	2

$$2 \text{ and } 5$$
$$\downarrow \qquad \downarrow$$
$$(2x + 3)(5x + 2)$$
$$\uparrow \qquad \uparrow$$
$$3 \text{ and } 2$$

$$(2x + 3)(5x + 2) = 10x^2 + 15x + 4x + 6 = 10x^2 + 19x + 6 \ ✔$$

There is no question that factoring a quadratic trinomial which has a leading coefficient other than 1 is the most difficult of all the cases. To help students understand how to factor polynomials of this form, you may want to show students the following whiteboard discussion.

$$10x^2 + 19x + 6$$

Class: Let's suppose that the expression above can be written as the partially blank product below.

$$(\underline{\hspace{1cm}} x + \underline{\hspace{1cm}})(\underline{\hspace{1cm}} x + \underline{\hspace{1cm}})$$

Let's guess that the two factors of 10 needed to fill in the coefficients of x are 2 and 5.

$$(\underline{\hspace{0.3cm}2\hspace{0.3cm}} x + \underline{\hspace{1cm}})(\underline{\hspace{0.3cm}5\hspace{0.3cm}} x + \underline{\hspace{1cm}})$$

Class: What two factors of 6 do you think will fill in the remaining two blanks?

Response: We could try 1 and 6. But we could also try 2 and 3.

Class: Try using 2 and 3. Some of you find (2x + 2)(5x + 3) and some of you find (2x + 3)(5x + 2).

Response: (2x + 2)(5x + 3) gives $10x^2 + 16x + 6$

(2x + 3)(5x + 2) gives $10x^2 + 19x + 6$ ✓

Notice that in the whiteboard discussion above, the teacher has already started the factorization with the correct factors of 10. Notice also that writing a blank form to fill in guesses helps keep the experimentation on track. If students fill in the two blanks incorrectly, it is an easy matter to erase the guesses and switch them around.

Students who participate in their learning are active learners. The following whiteboard display shows a slightly modified version of the whiteboard discussion above. The purpose of these whiteboard presentations is to encourage teachers to develop creative ways of teaching algebra.

$$10x^2 + 19x + 6$$

Class: Let's suppose that the expression above can be factored. Let's guess that the two factors of 10 needed to fill in the coefficients of x are 2 and 5.

$$(\underline{2}x + \underline{\quad})(\,\underline{5}x + \underline{\quad})$$

Class: Try using 2 and 3. Maggie, come to the whiteboard and fill in the blanks at the left below with 2 and 3. Greg, come to the whiteboard and fill in the blanks at the right below with 3 and 2.

Maggie   Greg

$(\underline{2}x + \underline{2})(\underline{5}x + \underline{3})$ $(\underline{2}x + \underline{3})(\underline{5}x + \underline{2})$

Class: Robert, find the product that Maggie has written.
Michelle, find the product that Greg has written.

Response: Robert: $(2x + 2)(5x + 3)$ gives $10x^2 + 16x + 6$

Michelle: $(2x + 3)(5x + 2)$ gives
$$10x^2 + 19x + 6 \checkmark$$

Class: Let's summarize what we have just done.

Another approach to factor $10x^2 + 19x + 6$ is by using a handout. The handout should provide the basic structure of the factoring problem, such as two parentheses and four blank lines as shown below.

$$10x^2 + 19x + 6 = (\underline{\quad}x + \underline{\quad})(\underline{\quad}x + \underline{\quad})$$

Students should work in small groups to facilitate discussion of possible solutions. We also recommend using handouts only after the students have seen the teacher model the process of factoring. Handouts should be used as independent practice, and the teacher may reserve their use for homework assignments.

Using Factoring to Solve Equations

One of the reasons for learning how to factor polynomial expressions is that factoring reveals the link between linear factors of expressions and roots of related equations. The importance of this link comes to light in more advanced mathematics courses. Nonetheless, in an introductory algebra course, students can begin to appreciate how linear factors of quadratic expressions are related to roots of quadratic equations.

Two quick examples will show how factoring a quadratic expression will give the roots of an equation. An important notion here is that a product equal to 0 must have a factor equal to 0.

$$\text{If } ab = 0, \text{ then } a = 0 \text{ or } b = 0.$$

Consider $t^2 - 5t + 6 = 0$.

$$\begin{aligned}
t^2 - 5t + 6 &= 0 \\
(t - 2)(t - 3) &= 0 \quad \text{since } t^2 - 5t + 6 = (t - 2)(t - 3) \\
t - 2 = 0 \text{ or } t - 3 &= 0 \\
t = 2 \text{ or } t &= 3
\end{aligned}$$

The solutions to $t^2 - 5t + 6 = 0$ are 2 and 3.

Consider $3t^2 - 8t + 5 = 0$.

$$\begin{aligned}
3t^2 - 8t + 5 &= 0 \\
(3t - 5)(t - 1) &= 0 \quad \text{since } 3t^2 - 8t + 5 = (3t - 5)(t - 1) \\
3t - 5 = 0 \text{ or } t - 1 &= 0 \\
3t = 5 \text{ or } t &= 1 \\
t = \frac{5}{3} \text{ or } t &= 1
\end{aligned}$$

The solutions to $3t^2 - 8t + 5 = 0$ are $\frac{5}{3}$ and 1.

Keeping an Algebra Journal

Students should be encouraged to keep a record of their factorization work in their algebra journals. For example, students can make annotations about things that are difficult to understand. Also, students should be encouraged to keep track of their successes. Students who are successful usually develop self-confidence and motivation to learn more.

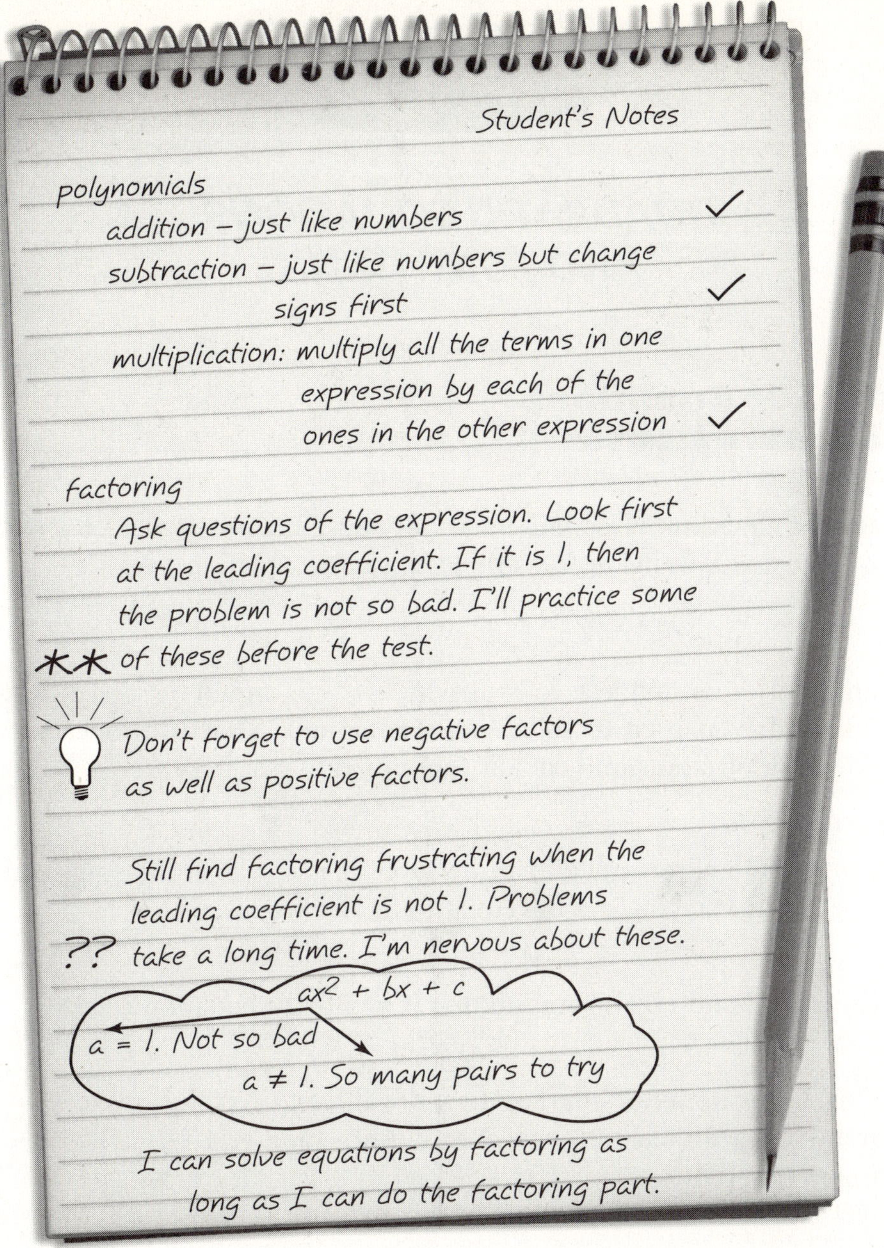

The importance of keeping a journal cannot be underestimated. Many students tend to focus their attention on getting a correct answer. With this mindset, they may not be inclined to step back and reflect on how they solve different problems through different approaches. The journal entries above suggest that the student has decided to be aware of what to look for when beginning a problem. Journal writing is an excellent way to develop the students' metacognitive skills, knowing how much they know about their own knowledge and what strategies they use when they learn something new.

QUADRATIC EQUATIONS AND FUNCTIONS

Quadratic Equations and Square Roots

A **quadratic equation** in x is an equation whose highest power of x is x^2. Each of the equations below is a simple quadratic equation.

$$x^2 = 4 \qquad x^2 = \frac{16}{25} \qquad x^2 = 5$$

Even an equation as simple as $x^2 = a$, where $a > 0$, cannot be solved by the same methods you used to solve linear equations. That is, properties of addition, multiplication, and equality are not enough.

Consider $x^2 = 4$.

Using trial-and-error, you can discover that if $x = 2$, the equation is true and that if $x = -2$, the equation is also true.

$$2^2 = 4 \; \checkmark \quad \text{and} \quad (-2)^2 = 4 \; \checkmark$$

Using trial-and-error, you find that if $x = 0$, the equation is false.

$$0^2 \neq 4$$

This discussion indicates that there are at least two solutions to $x^2 = 4$ and that not all real numbers are solutions to $x^2 = 4$.

Consider $x^2 = \frac{16}{25}$.

The numerator, 16, of the fraction can be written as $4 \cdot 4$ and the denominator, 25, of the fraction can be written as $5 \cdot 5$. Therefore, you can write the product below.

$$\frac{4}{5} \cdot \frac{4}{5} = \frac{16}{25}$$

This product tells you that $\frac{4}{5}$ is the square root of $\frac{16}{25}$.

You can also write a product involving two negative numbers as below.

$$\left(-\frac{4}{5}\right)\left(-\frac{4}{5}\right) = \frac{(-4)(-4)}{(5)(5)} = \frac{16}{25}$$

Again, the product tells you that $-\frac{4}{5}$ is the square root of $\frac{16}{25}$.

Therefore, the equation $x^2 = \frac{16}{25}$ has $\frac{4}{5}$ and $-\frac{4}{5}$ as its solutions.

The following whiteboard presentation can help students understand square roots better.

> **Class:** Look at the equation below. Use trial-and-error to think of any numbers that satisfy the equation. What are they?
>
> $$x^2 = 121$$
>
> **Response:** The number 11 works. $11 \cdot 11 = 121$ ✓
>
> **Class:** Any others?
>
> **Response:** The number −11 works. $(-11) \cdot (-11) = 121$ ✓
>
> **Class:** Do you notice any relationship between the solutions you found?
>
> **Response:** They are opposites of one another.

What students discover is true. If a is a positive number, then $x^2 = a$ has two solutions and they are opposites of one another.

Not all solutions to quadratic equations are rational numbers.

Consider $x^2 = 5$.

Using trial-and-error, you will never find any integers or rational numbers that are solutions to $x^2 = 5$. The following analysis in table form suggests that there may be at least one number that satisfies $x^2 = 5$ but that number may not be a rational number.

x	2	2.2	2.23	2.236	2.23605
x^2	4	4.84	4.9729	4.999696	4.999919603

If you continue the approximation process, you will find that the decimal number at right gives a close approximation to a solution.

$$2.236067977$$

---

quadratic equation: An equation of the form $y = ax^2 + bx + c$, where a, b, and c are real numbers and $a \neq 0$.

We now reflect on an important difference between linear equations in one variable and quadratic equations in one variable.

- If you are given a linear equation in one variable containing numbers that are rational and not all real numbers are solutions, the solution will be rational.

- However, even a simple quadratic equation in one variable involving rational numbers may have solutions that are not rational numbers. The solutions to $x^2 = 4$ and to $x^2 = \frac{16}{25}$ are rational numbers. The solutions to $x^2 = 5$ are not rational numbers.

perfect square: A number whose square roots are rational numbers.

A **perfect square** is any rational number whose square roots are rational numbers. The number $\frac{16}{25}$ is a perfect square. The solutions to $x^2 = \frac{16}{25}$ are denoted as $\sqrt{\frac{16}{25}}$ and $-\sqrt{\frac{16}{25}}$. Since these numbers can be written as $\frac{4}{5}$ and $-\frac{4}{5}$, the radical symbol $\sqrt{}$ is dropped. The solutions to $x^2 = 5$ are denoted $\sqrt{5}$ and $-\sqrt{5}$. Since 5 is not a perfect square, the radical symbol $\sqrt{}$ may not be dropped.

The following fact gives important information about $x^2 = a$, where $a > 0$.

> ### Solutions to $x^2 = a$, where $a > 0$
>
> If a is a perfect square, then the solutions to $x^2 = a$ are rational numbers.
>
> If a is not a perfect square, then the solutions to $x^2 = a$ are not rational numbers. However, you can use a calculator to approximate them.

The study of solutions to $x^2 = a$, where $a > 0$, gives rise to the definition of square root.

square root: If a is a number greater than or equal to zero, $\sqrt{a}$ represents the positive, or principal, square root of a and $-\sqrt{a}$ represents the negative square root of a.

> ### Definition of Square Root
>
> Let $x^2 = a$, where $a > 0$. If x is a solution to $x^2 = a$, then x is called a **square root** of a.

$\sqrt{a}$ denotes the positive square root of a and $-\sqrt{a}$ denotes the negative square root of a.

Finding the square of a number and finding its square root are operations that are inverses of one another.

$$5^2 = 25 \quad \text{and} \quad \sqrt{25} = 5$$

To carry out these two operations on a calculator, look for calculator keys that might look like these.

square a number: $\boxed{x^2}$ take the square root: $\boxed{\text{2nd}}$ $\boxed{x^2}$

For example, to evaluate $\sqrt{265}$, consider the illustration at right. Pressing $\boxed{\text{2nd}}$ makes the calculator carry out the inverse of the operation on the key labeled $\boxed{x^2}$.

$\boxed{\text{2nd}}$ $\boxed{x^2}$ *take the square root*

265

$\boxed{\text{ENTER}}$ *calculate*

One aspect of number sense is having an idea of the magnitude of numbers. For example, to determine how large the positive solution to $x^2 = 265$ is, consider $\sqrt{265}$.

The table below shows values of x and x^2.

x	13	14	15	16	17	18	19
x^2	169	196	225	256	289	324	361

Since 265 is between 256 and 289, you may conclude that $16 < \sqrt{265} < 17$. This means that the positive solution to $x^2 = 265$ is somewhere between 16 and 17. This gives you some number sense about the size of $\sqrt{265}$. The numbers in the table above also suggest that the larger a positive number is, the larger its positive square root will be.

Our discussion so far allows you to write exact solutions to equations of the form $x^2 = a$, where $a > 0$.

$$\begin{array}{ll} \text{equation} & \text{solutions} \\ x^2 = 100 & \rightarrow \quad 10 \text{ and } -10 \\ x^2 = \dfrac{49}{64} & \rightarrow \quad \dfrac{7}{8} \text{ and } -\dfrac{7}{8} \\ x^2 = 17 & \rightarrow \quad \sqrt{17} \text{ and } -\sqrt{17} \end{array}$$

Equations of the form $x^2 = a$, where $a > 0$, are solved by taking the square root of each side of the equation. Besides the properties of equality, students may now take the square root of each side of an equation as a strategy for solving quadratic equations.

When students learn to solve $x^2 = a$, where $a > 0$, perhaps the most important fact that students should not forget is that the equation has two solutions, and that the solutions are opposites of one another. For example, if students see the equation $x^2 = 36$, they may be quick to report 6 as a solution and forget that -6 is also a solution.

In this discussion, notice that the requirement that a be positive is important. It has been stated many times. Suppose that a were a negative number. For example, consider $x^2 = -4$.

- If you pick a positive number for x, then you will have the following product.

  (positive number x)(positive number x) = positive number

  That is, the product will be positive, not negative.

- If you pick a negative number for x, then you will have the following product.

  (negative number x)(negative number x) = positive number

The product again will be positive, not negative.

In other words, you will never find any real number whose square is a negative number. Therefore, to have solutions to $x^2 = a$, it is necessary that a be positive.

Sometimes problems have important restrictions on the type of solution required. For example, consider the square at right. How long is each side, s, of the square?

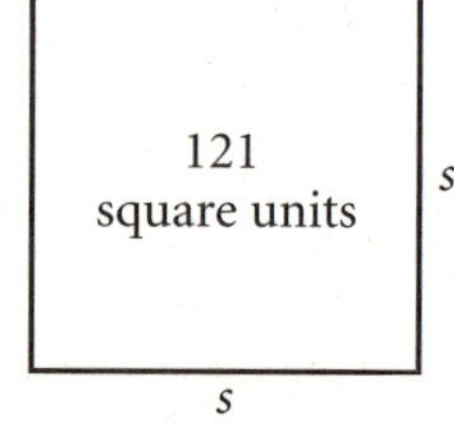

The area of the square is 121 square units. According to the formula from geometry, the area of a square is the square of the length of one side.

The equation $s^2 = 12$, where s represents the length of a side of the square, can be used to answer the question. According to our previous discussion on square roots, the solutions to $s^2 = 12$ are 11 and -11.

$$(11)(11) = 121 \ \checkmark \qquad (-11)(-11) = 121 \ \checkmark$$

However, only the positive solution, 11, makes sense in the context of this problem because length must be nonnegative. Therefore, the square is 11 units on a side.

Solving a Quadratic Equation That Has No Linear Term

As stated before, it is easy to write down the solutions to an equation of the form $x^2 = a$, where $a > 0$. Simply write the positive and negative square roots of a. Then take the square roots, if the number a is a perfect square, or leave the square root sign if the number a is not a perfect square.

In $25x^2 - 4 = 0$, x^2 is multiplied by a number, then 4 is subtracted. The result is 0. This equation is slightly different from $x^2 = 4$. Suppose you want to find the solutions to $25x^2 - 4 = 0$. To do this, you must take some preliminary steps.

Consider $25x^2 - 4 = 0$.

You can apply properties of equality to obtain an equivalent equation that has the form $x^2 = a$, where $a > 0$.

$$25x^2 - 4 = 0$$
$$25x^2 = 4$$
$$x^2 = \frac{4}{25} \quad \leftarrow \textit{This equation has the form } x^2 = a.$$

Now examine the equivalent equation $x^2 = \frac{4}{25}$. The left side of the equation is x^2 and the right side of the equation is a positive number. Now you can take the square root of each side to get the solutions.

$$x = \sqrt{\frac{4}{25}} \text{ or } -\sqrt{\frac{4}{25}}, \text{ that is } x = \frac{2}{5} \text{ or } -\frac{2}{5}$$

Therefore, the solutions to $25x^2 - 4 = 0$ are $\frac{2}{5}$ and $-\frac{2}{5}$.

You can write the solutions to $25x^2 - 4 = 0$ as $\pm\frac{2}{5}$. The symbol $\pm$, which is read "plus or minus," indicates that both the positive and negative square roots are intended. What happens if the equation $25x^2 - 4 = 0$ is replaced by an equation in which the number on the right side is not 0? Can such an equation be solved? If so, what method should we use?

Consider $25x^2 + 4 = 40$.

As before, you can apply properties of equality to obtain an equivalent equation that has the form $x^2 = a$, where $a > 0$.

$$25x^2 + 4 = 40$$
$$25x^2 = 36$$
$$x^2 = \frac{36}{25} \quad \leftarrow \quad \textit{This equation has the form } x^2 = a, \textit{ where } a > 0$$
$$x = \pm\sqrt{\frac{36}{25}} = \pm\frac{6}{5}$$

Therefore, the solutions to $25x^2 + 4 = 40$ are $\pm\frac{6}{5}$.

Our discussion about solving quadratic equations has moved from simplest to more complex cases. The following diagram shows what we have learned about solving quadratic equations.

Learn how to solve $x^2 = 4$

Extend and adapt.

Learn how to solve $25x^2 - 4 = 0$

Extend and adapt.

Learn how to solve $25x^2 + 4 = 40$

Notice that once students gain experience in solving one type of equation, they can transfer their skills to solve a broader class of equations. Transfer of knowledge is a good indicator of understanding and meaningful learning.

To solve $25x^2 + 4 = 40$ at the whiteboard, you may want to consider the following strategy. Give students the equation $25x^2 + 4 = 40$ and ask them what they would do in order to get an equation of the form

$$x^2 = \text{positive number}.$$

To guide students' thinking, you can ask them questions such as, What properties of equality do you think will be helpful? or How do you know you have gotten the desired equation?

Class: Look at the equation below. What would you do to get an equation like the one below it?

$$25x^2 + 4 = 40$$

$$x^2 = \text{positive number}$$

Response: Start by getting rid of the 4 on the left side of the equation.

$$25x^2 = 36$$

Class: Now what would you do?

Response: Divide each side by 25 to get rid of it.

$$x^2 = \frac{36}{25}$$

Class: How close to being finished are you?
Response: Just take square roots and it will be done.

$$x = \pm\frac{6}{5}$$

As mentioned earlier in this book, algebra and geometry are related disciplines. In an introductory algebra course, students learn about the Pythagorean Theorem. This theorem provides an important relationship among the lengths of the sides of a right triangle. Recall that a right triangle is a triangle that has one angle measuring $90°$.

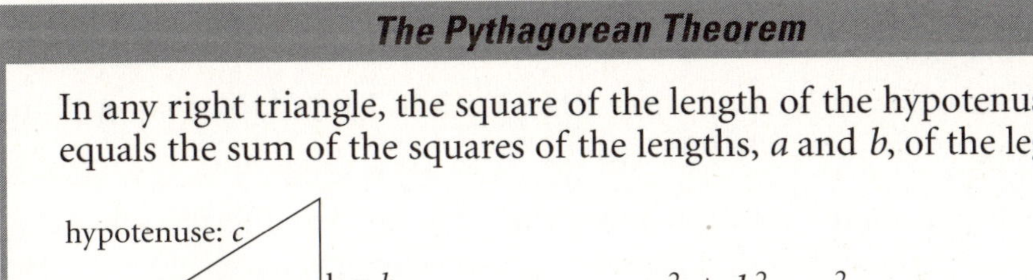

If you know a triangle is a right triangle and you know the lengths of any two sides in it, you can use algebra to find the length of the third side. To do this, you write and solve a quadratic equation.

Consider this application problem.

> Suppose that a maintenance crew wants to install a support wire for a new transmission tower. When built, the tower will be 400 feet tall. Four support wires, all equal in length, will be anchored 85 feet from the base. Find the length of each support wire to the nearest foot.

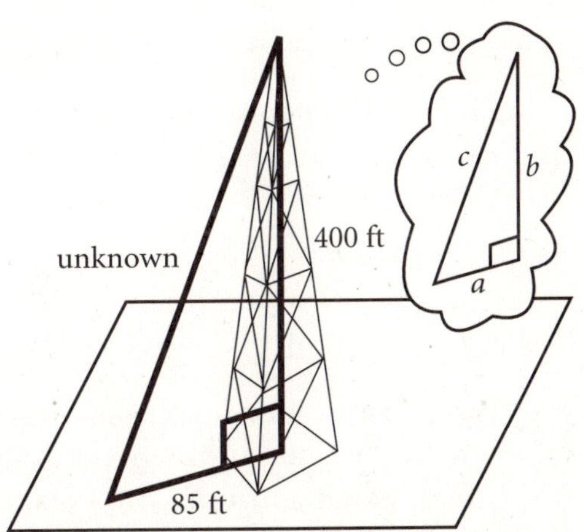

You can represent this verbal problem visually. For example, students can sketch the diagram at right on the whiteboard or on paper. Notice that the diagram clearly shows the right triangle and the right angle symbol. The known lengths are labeled and placed where they should be. The word "unknown" is placed where it should be. With a diagram that represents the information given and the unknown, you can write and solve an equation.

$$c^2 = 400^2 + 85^2$$

$$c = \sqrt{400^2 + 85^2} \quad \leftarrow \textit{Reject the negative root.}$$

Since the unknown length should be calculated to the nearest foot, you will need a calculator to complete the solution. To evaluate $\sqrt{400^2 + 85^2}$ on a calculator, you need both the squaring function and the square root function. Notice in the display below that you must use parentheses to group the expression under the square root symbol.

| 2nd | | x^2 | | (| 400 | x^2 | | + | 85 | x^2 | |) | | ENTER |

The calculator screen will show 408.9315346. To find the answer to the nearest foot, round 408.9315346 to 409.

Notice in the calculator solution above that students need to use what they learned about the order of operations to get the correct answer. That is, they must use parentheses to group the quantity whose square root they need to find.

All of the steps taken to solve $25x^2 + 4 = 40$ can be carried out to solve the general equation $ax^2 + c = d$. When the steps are carried out, you will get the formula stated below. This formula will give the solutions.

$$x = \pm\sqrt{\frac{d - c}{a}}$$

Students may ask you this question.

If I can apply a formula to solve all equations that have the form $ax^2 + c = d$, why not memorize the formula and simply use it when I need it?

In $25x^2 + 4 = 40$, $a = 25$, $c = 4$, and $d = 40$. So, $x = \pm\sqrt{\frac{40 - 4}{25}}$. I can use a calculator to find $\sqrt{\frac{40 - 4}{25}}$. So, why go through all the solving steps? To answer these questions, we encourage you to think about your role as a mathematics teacher. If our goal is to teach mathematics for understanding, we should assign as much importance to the teaching of procedures and formulas as to the teaching of concepts, reasoning, and logic. Overusing formulas may make students forget about all the important steps in the reasoning. Consequently, they will not gain understanding of the mathematical reasoning that is connected to a formula. Rather, students may become adept at using formulas mechanically.

When students learn methods for solving linear equations in one variable, they are taught how to solve all kinds of linear equations. For example, students are expected to solve any of the equations below.

$$2x = 4 \qquad\qquad 3(d - 3) = 5$$
$$3t + 5 = 6t + 1 \qquad -(y + 1) - 4 = 3(y + 3) - 7$$

Another agenda in an introductory algebra course is solving different kinds of quadratic equations. By the time students finish an introductory algebra course, they are expected to be able to solve various kinds of linear equations and quadratic equations in one variable. For example, students should be able to solve any of the equations below.

$$x^2 = 144 \quad x^2 = 110 \quad\quad 3x^2 = 75 \quad\quad 3x^2 = 9$$

$$2x^2 + 10 = 82 \quad\quad 7x^2 - 5 = 144$$

Even an equation like $5x^2 + 4 = 3x^2 + 36$ should not be too difficult to solve at this point in the students' learning. The fact that the variable is on both sides of the equation should not be an obstacle. On the contrary, this should represent a chance for students to transfer their knowledge to a new situation. With guidance, students should be able to proceed as shown below.

$$5x^2 + 4 = 3x^2 + 36$$

$$2x^2 + 4 = 36 \quad \leftarrow \textit{Subtract } 3x^2 \textit{ from each side.}$$

$$2x^2 = 32 \quad \leftarrow \textit{Subtract 4 from each side.}$$

$$x^2 = 16 \quad \leftarrow \textit{Divide each side by 2.}$$

$$x = \pm 4$$

Using the Zero-Product Property

Recall that the goal in solving an equation is to arrive at an equation that is in the following form.

$$\text{variable name} = \text{number solution(s)}$$

In chapter 5, the Zero-Product Property was discussed in the context of solving equations. We now review factoring as a method to solve quadratic equations.

Consider $x^2 + 8x + 12 = 0$.

To solve this equation by using the Zero-Product Property, you need to take three steps.

Step ❶: Factor.

Since $2 + 6 = 8$ and $(2)(6) = 12$, then $x^2 + 8x + 12 = (x + 6)(x + 2)$.

Step ❷: Write linear equations.

$(x + 6) = 0$ and $(x + 2) = 0$

Step ❸: Solve linear equations using properties of equality.

$x = -6$ and $x = -2$

$$x^2 + 8x + 12 = 0$$

$\downarrow$ *factoring skills*

$$(x + 6)(x + 2) = 0$$

writing two linear equations

$$x + 6 = 0 \qquad\qquad x + 2 = 0$$

solving linear equations

$$x = -6 \qquad\qquad x = -2$$

variable = number   *variable = number*

The strategy worked. The solutions are -6 and -2.

Students in an introductory algebra course should be able to solve a quadratic equation by factoring and setting factors equal to 0. Solving a quadratic equation by factoring helps to remind students that the skills they learned earlier will return in new contexts and for new purposes.

After completing lessons in which factoring is used to solve equations, students can now expand the range of equations they can solve. For example, students should be able to solve all of the equations below.

$$x^2 = 144 \qquad x^2 = 110 \qquad 3x^2 = 75 \qquad 3x^2 = 9$$

$$2x^2 + 10 = 82 \qquad 7x^2 - 5 = 144$$

$$x^2 - 3x = 0 \qquad x^2 + 5x + 6 = 0 \qquad 10x^2 + 19x + 6 = 0$$

Solving an Area Problem

Mathematical skills and methods in an introductory algebra course are usually accompanied by real-world applications.

Consider the rectangle shown below. What are its dimensions?

$$2x + 3$$

$x + 5$ | 130 square units

This geometry problem provides an opportunity to practice problem-solving skills related to solving equations.

A whiteboard illustration shows one possible way to present this problem to students.

Class: How can we find the dimensions of the rectangle at right?

$2x + 3$

$x + 5$ | 130 square units

Response: We know that length times width gives area.
$$(2x + 3)(x + 5) = 130$$
Does that help, teacher?

Class: Let's find out. Perform the multiplication and then subtract 130 from both sides of the equation. Then let's think about what we have.
$$2x^2 + 13x - 115 = 0$$

As students carry out procedures, ask them to verbally state the purpose of solving the equation in this problem. Sometimes students lose sight of their goal while they are in the middle of solving a problem.

Class: Suppose you can solve the equation for x. What will the solutions tell you?

Response: Put solutions into $x + 5$, the expression for width, and $2x + 3$, the expression for length, to get numbers for the dimensions of the rectangle.

Class: Very good! Now think about how to find x.

Response: We could try factoring. But, that looks tough because of the number 115.

Class: Try looking for factors of 115.

Response: We think we found a factorization.
$$(x - 5)(2x + 23) = 0$$

Class: Follow through and see what you get for x.

Response: $\qquad x = 5$ or $x = -\dfrac{23}{2}$

Reject $-\dfrac{23}{2}$ since it makes $x + 5$ negative.
The dimensions are 10 units and 13 units.

The following flowchart shows the thinking underlying the solution.

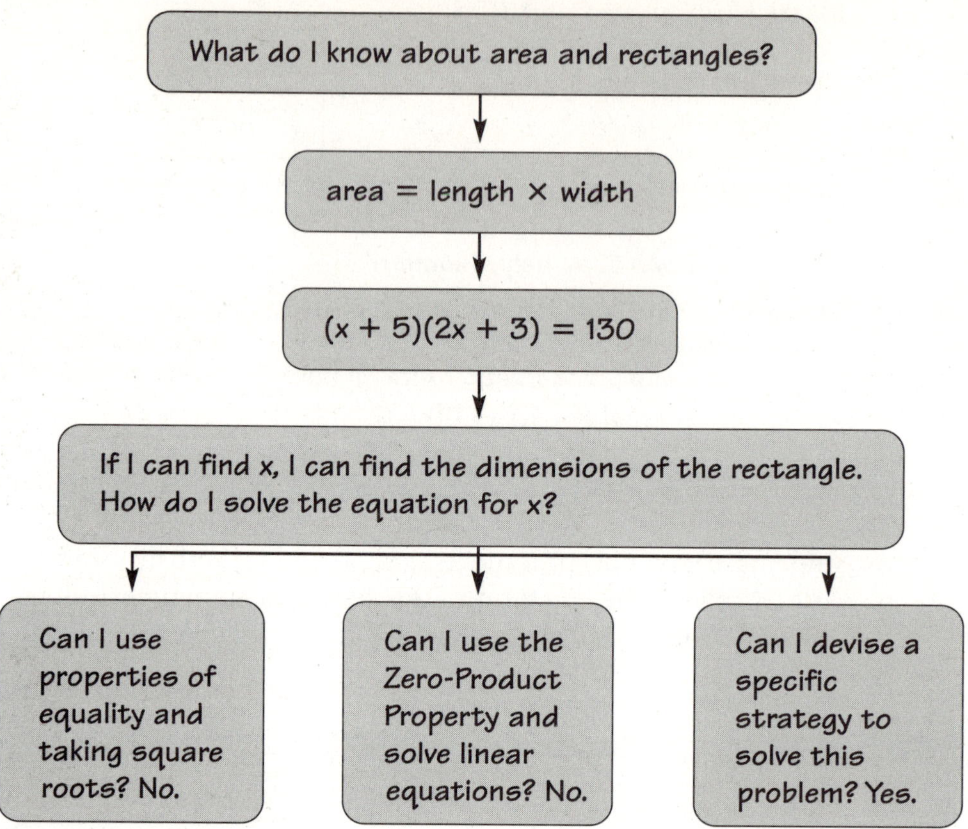

Students in an introductory algebra course gradually work up to a problem as complicated as the one just presented. This underscores the following facts.

Building Mathematical Skills

- Mathematical skills build on one another. Therefore, students need to have confidence that they are learning skills before they move on to other skills.

- As students acquire more skills, their ability to solve more complex problems increases.

The Quadratic Formula

One of the objectives in an algebra course is the development of problem-solving skills. The diagram below shows the learning behaviors algebra students must acquire in order to become successful independent problem solvers.

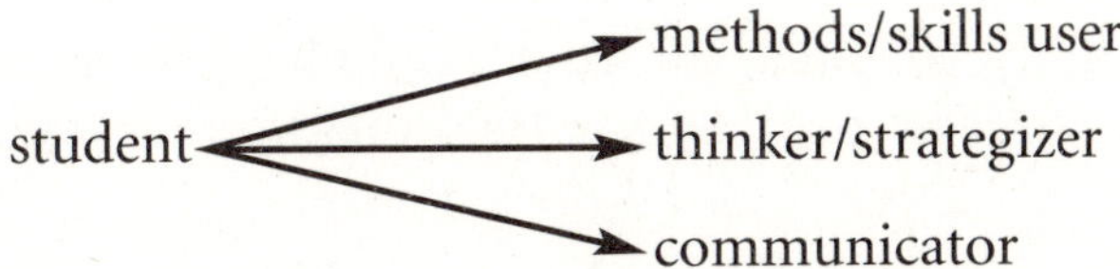

However, when existing and available methods are not sufficiently adequate, problem solvers require new methods. For example, students will not be able to factor $t^2 + 2t - 5$ by using integers in the factors. Therefore, they will not be able to use the Zero-Product Property in any convenient way to solve $t^2 + 2t - 5 = 0$. Consequently, it becomes necessary to invent new solution methods.

Using logical reasoning, mathematicians have been able to derive a formula that will give all solutions to all quadratic equations. The formula is called the *quadratic formula* and it is stated below.

The Quadratic Formula

If $ax^2 + bx + c = 0$, where $a \neq 0$, then $x = \dfrac{-b \pm \sqrt{b^2 - 4ac}}{2a}$.

Shown below is an example of how to apply the quadratic formula.

Consider $t^2 + 2t - 5 = 0$.

$$t^2 + 2t - 5 = 0$$

$$t = \frac{-2 \pm \sqrt{2^2 - 4(1)(-5)}}{2(1)} \qquad \leftarrow \text{ Apply the quadratic formula with } a = 1, b = 2, \text{ and } c = -5.$$

$$t = \frac{-2 \pm \sqrt{24}}{2}$$

$$t = \frac{-2 \pm 2\sqrt{6}}{2} \qquad \leftarrow \sqrt{24} = 2\sqrt{6}$$

$$t = -1 \pm \sqrt{6}$$

The solutions to $t^2 + 2t - 5 = 0$ are $-1 - \sqrt{6}$ and $-1 + \sqrt{6}$.

Some students may look at the equation $t^2 = -2t + 5$, recognize it to be a quadratic equation, and incorrectly label the coefficients as $a = 1$, $b = -1$, and $c = 5$. Students must rewrite this equation in standard form before identifying a, b, and c.

$$t^2 = -2t + 5$$
$$t^2 + 2t - 5 = 0 \qquad \leftarrow \qquad a = 1, b = 2, \text{ and } c = -5.$$

To use the quadratic formula, students should be able to read and evaluate a formula, take square roots, and write a quadratic equation in standard form.

However, for many quadratic equations the quadratic formula is not the most efficient solution method. For example, students should not be encouraged to use the quadratic formula to solve a quadratic equation like $x^2 = 4$. Instead, students should be given meaningful practice so they can choose between different solution methods.

After students learn how to use the quadratic formula, they should learn about the discriminant of a quadratic equation. The **discriminant** of $ax^2 + bx + c = 0$ is $b^2 - 4ac$. You can have students discover how to use the discriminant through the following whiteboard activity.

discriminant: For a quadratic equation of the form $ax^2 + bx + c = 0$, where $a \neq 0$, the discriminant is $b^2 - 4ac$.

Class: Part of the quadratic formula is below.

$$\sqrt{6^2 - 4ac}$$

What can you conclude if b2 – 4ac is positive?

Response: You may take the square root. You get two values for solutions.

Class: What can you conclude if b2 – 4ac equals 0?
Response: The square root equals 0. That part of the formula goes away. There is one answer.

Class: What can you conclude if b2 – 4ac is negative?
Response: You get a negative number under the square root symbol. There are no real solutions.

Teachers should look for creative ways to involve students actively in the learning process. In a chapter on quadratic equations, students learn many different methods for solving equations. Students also need to know when to use a particular method. To invite students to practice choosing different methods for different quadratic equations, consider the following activity.

Prepare several index cards, each having one quadratic equation on it. Label the cards A, B, C, and so on. Prepare several index cards with phrases that identify solution methods or parts of solution methods. Label these cards 1, 2, 3, and so on.

Some examples of equation cards as well as of method cards are shown below. Depending on how many equation cards and method cards you make, you may need to make multiple copies of methods cards.

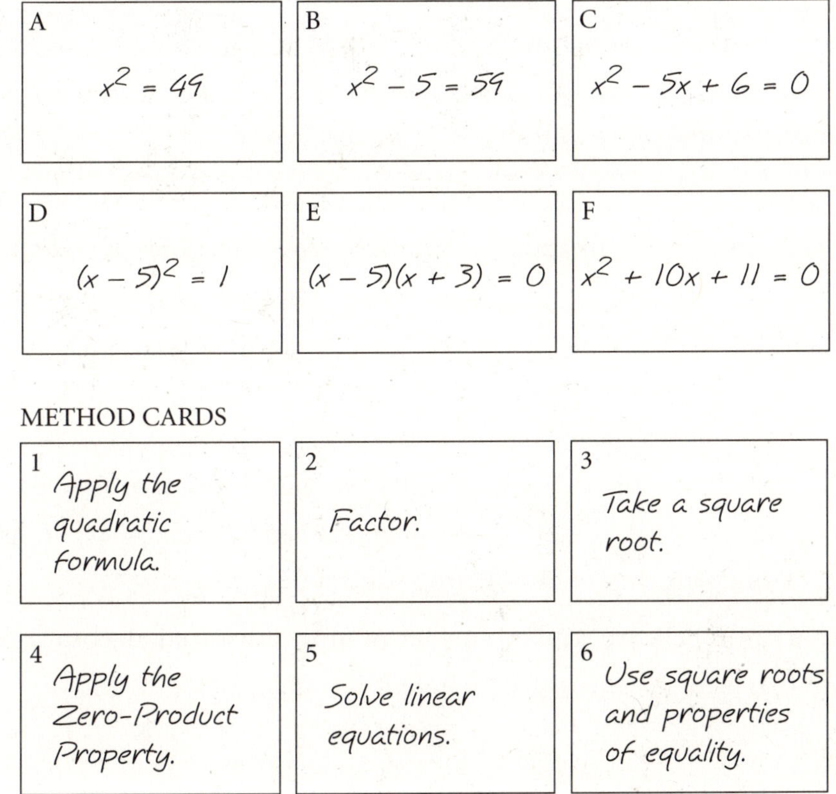

Hand out the cards labeled A, B, C, and so on. Then randomly hand out the cards labeled 1, 2, 3, and so on. Students then seek to match the equation cards with the method cards.

Students can work in small groups. Through this activity, students will continue to learn to read an equation, observe what is in it, make choices based on what they read and see, and explain to group members why some choices are better than others.

Students should also be taught that not every quadratic equation has real solutions. For example, if $x^2 + 1 = 0$, then $x^2 = -1$. There is no real number whose square is negative. To solve this conflict, mathematicians chose to expand the world of numbers from the set of real numbers to include what are called *imaginary numbers* so that all quadratic equations would have solutions in the new set of numbers.

Imaginary Numbers

The imaginary number i is defined by $i = \sqrt{-1}$. The number i is called the *imaginary unit*. If $r > 0$, then $\sqrt{-r} = i\sqrt{r}$.

With the creation of imaginary numbers, you can solve any quadratic equation.

Consider $z^2 + 4 = 0$.

$$z^2 + 4 = 0$$
$$z^2 = -4$$
$$z = \pm\sqrt{4} \qquad \leftarrow \textit{Take the square root of each side.}$$
$$z = \pm 2i \qquad \leftarrow \textit{Apply the definition of } i = \sqrt{-1}.$$

The solutions to $z^2 + 4 = 0$ are $2i$ and $-2i$.

Imaginary numbers are studied further in more advanced algebra courses.

Quadratic Functions

The equation $y = x^2$ defines a function because for each value of x there is exactly one square of x. A table of values can give students an idea of the graph of $y = x^2$.

x	-3	-2	-1	0	1	2	3
y	9	4	1	0	1	4	9

When you graph the ordered pairs in the table and draw a smooth curve through them, you will get the graph shown below. The characteristics of this graph are summarized below.

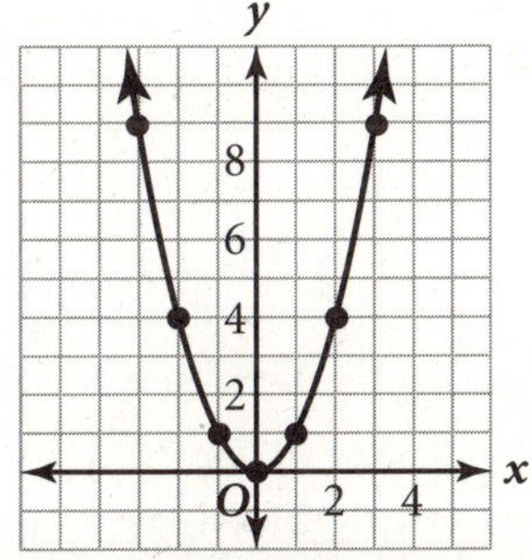

Characteristics of the Graph of $y = x^2$

- There is no straight line that will go through all the points. The graph of $y = x^2$ is not like the graph of a linear function.

- Scanning the graph from left to right, you see the graph decreasing, passing through $(0, 0)$, and then increasing.

- The y-axis acts as a mirror. For each point on the graph and to the left of the y-axis, there is a companion point to the right of the y-axis.

- The graph is continuous, smooth, and extends indefinitely up to the left and up to the right, as indicated by the arrowheads.

The characteristics of the graph of $y = x^2$ give rise to vocabulary terms that are relevant to graphs of quadratic functions.

Terminology of Graphs of Quadratic Functions

- The graph has a **U** shape called a **parabola**.

- The y-axis, which serves as a mirror, is called the **axis of symmetry** of the graph. Notice that $(-3, 9)$ and $(-2, 4)$ are on the graph of $y = x^2$. Their mirror images are $(3, 9)$ and $(2, 4)$, respectively.

- For $y = x^2$, the origin is the **vertex** of the parabola. It is the point that is both on the graph and on the axis of symmetry. Notice in the diagram above that $(0, 0)$ is its own mirror image.

parabola: The shape of the graph of a quadratic function. A parabola is defined as the set of all points that are the same distance from a fixed point, called the focus, and a fixed line, called the directix.

axis of symmetry: A line drawn through the graph of a function that divides the graph into two halves that are reflections of each other.

vertex: The point where a parabola changes direction.

The following examples illustrate the use of quadratic equations in real-world situations.

Example 1

Motorists know that the distance needed to stop a moving vehicle increases as motor vehicle speed increases. Experienced motorists know that stopping distance is not a linear function of speed. In other words, if speed is doubled, stopping distance is multiplied by 4. If speed is tripled, stopping distance is multiplied by 9. This information is represented in the diagram below, where s is speed and d is distance. Such information is crucial to assure adequate distance between vehicles to prevent accidents.

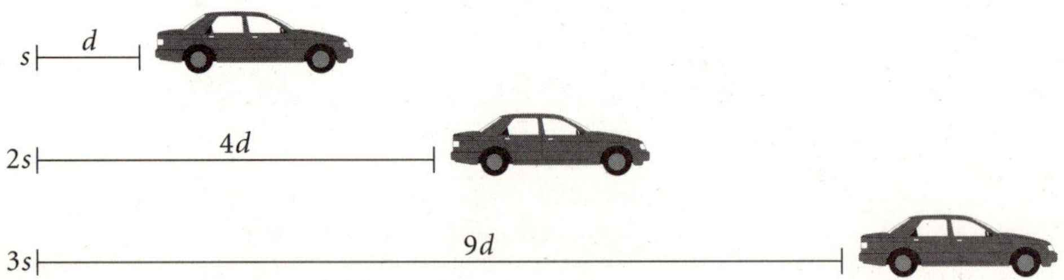

The equation at right gives a relationship between traveling speed, s, in miles per hour and stopping distance, d, in feet. The equation is a quadratic function.

$$d = \frac{2}{55}s^2$$

Example 2

Consider tanks A, B, and C having the same height. The radius of tank B, however, is twice that of tank A and the radius of tank C is three times that of tank A.

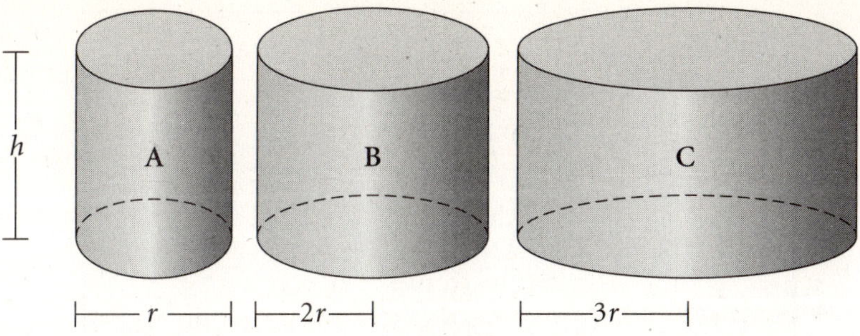

The formula for the volume, V, of a cylinder with radius r and height h is given at right. Using the formula, you can reason that the volumes of tanks B and C are 4 times and 9 times that of tank A, respectively.

$$V = \pi r^2$$

$$
\begin{array}{ccc}
\text{Tank A} & \text{Tank B} & \text{Tank C} \\[4pt]
V = \pi r^2 & V = \pi(2r)^2 & V = \pi(3r)^2 \\
 & = \pi(4r^2) & = \pi(9r^2) \\
 & = 4\pi r^2 & = 9\pi r^2
\end{array}
$$

For example, suppose that tank A contains 1000 cubic feet of water, tanks B contains 4000, and tank C contains 9000. Since water weighs about 62.3 pounds per cubic foot, the water in tank A will weigh 62,300 pounds, or 31.15 tons, the water in tank B will weigh 249,200 pounds, or 124.6 tons, and the water in tank C will weigh 560,700 pounds, or 280.35 tons.

Any equation of the form $y = ax^2 + bx + c$, where $a \neq 0$, is called a *quadratic function*. Its graph has a U shape similar to the graph of $y = x^2$. The graph also has an axis of symmetry. However, the y-axis is not necessarily the axis of symmetry anymore. Instead, the x-coordinate of the vertex, $-\dfrac{b}{2a}$, is the axis of symmetry.

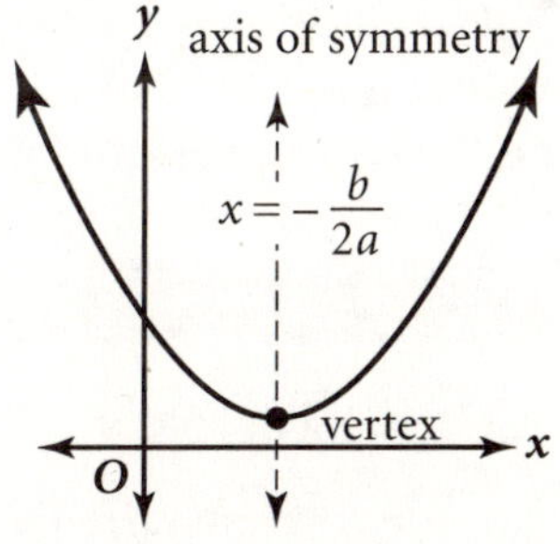

In an introductory algebra course, students learn some basic techniques for graphing such functions. In the discussion that follows, you will see that graphing a linear equation and graphing a quadratic equation involve different strategies.

Shown on the following page is an illustration of a worksheet that students can use to help guide them through the process of graphing a quadratic function such as $y = x^2 - 4x + 3$.

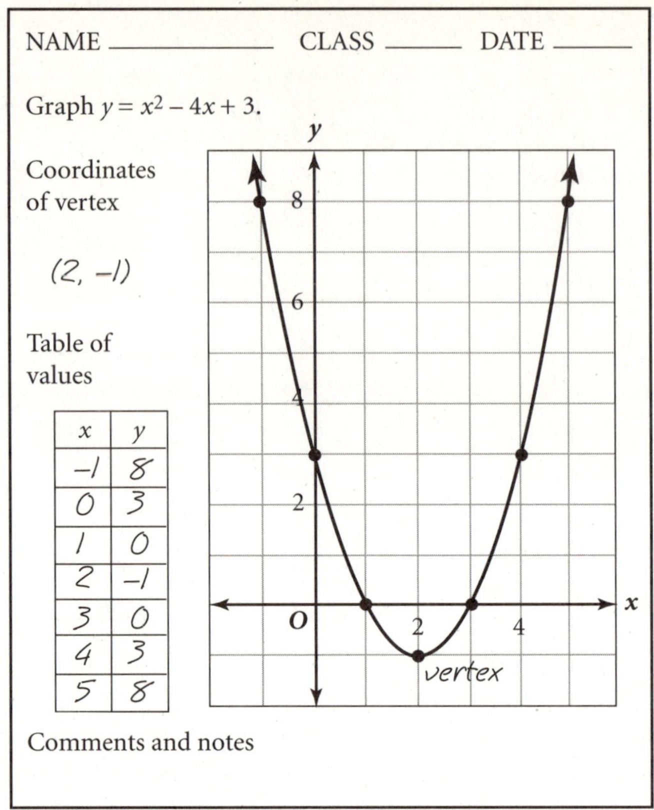

Before students are given independent practice, such as worksheets, the teacher should model the process of graphing a quadratic equation. For example, the teacher can hand out copies of the steps in the box below and students can read the steps as the teacher demonstrates the actual work on the whiteboard.

Quadratic Equations and Graphs

❶ Use the formula $-\dfrac{b}{2a}$ to find the x-coordinate of the vertex. Then find the corresponding value of y.

❷ Make a table of values using three values of x less than the x-coordinate of the vertex, the x-coordinate of the vertex, and three values of x greater than the x-coordinate of the vertex.

❸ Plot the seven ordered pairs from the table and draw a smooth curve through them. Arrowheads should be on the ends of the curve.

❹ Discuss the work and write any comments about graphing a quadratic function. For example, students might observe that by finding the coordinates of the vertex first, they have pinned down the highest (lowest) point on the graph as a major first step.

Keeping an Algebra Journal

A study of quadratic equations is of great importance in an introductory algebra course. As noted before, students should be encouraged to **summarize main points about quadratic equations in their journals.**

Student's Notes

Quadratic equations
  learn to take square roots of numbers ✓
  learn to take square roots to solve ✓
    simple quadratic equations

  use special techniques
    Zero-Product Property ✓
    completing the square ✓
✳✳ Memorize quadratic formula.
✳

$$x = \frac{-b \pm \sqrt{b^2 - 4ac}}{2a}$$

Remember the discriminant $b^2 - 4ac$

Practice exercises of different sorts
to practice choosing a method for
solution.

Quadratic functions
  U shape is the shape to look for.
  axis of symmetry cuts graph at vertex.

✳✳ x-coordinate of vertex is $\dfrac{-b}{2a}$
✳

  Make table of values with vertex in it.
  Sketch smooth curve through points.

RATIONAL EXPRESSIONS AND FUNCTIONS

Rational Numbers

The diagram at right may be used to help introduce students to the concept of fractions. Students should learn that a fraction represents a part of a whole. This diagram illustrates the fraction $\frac{5}{12}$. Later, students expand the concept to include fractions with the numerator greater than the denominator. The fraction $\frac{12}{5}$ is an example. This fraction is represented in the diagram below. Students learn to write $\frac{12}{5}$ as $2\frac{2}{5}$ and call $2\frac{2}{5}$ a mixed number.

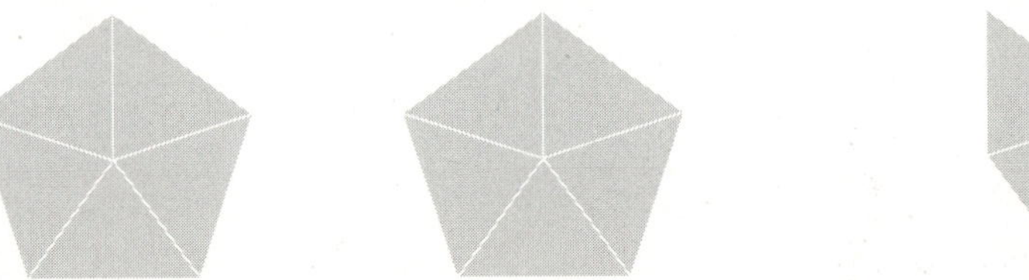

The study of fractions is an important foundation for the study of rational numbers. A **rational number** is any number that can be written in the form $\frac{a}{b}$, where a and b are integers and $b \neq 0$. Examples of rational numbers are $-\frac{5}{12}, \frac{5}{12}, 0, -13,$ and 7. Notice that 0 is a rational number because 0 divided by any nonzero number still equals 0. Similarly, -13 and 7 are rational numbers because each can be divided by 1. The table at right summarizes various types of rational numbers with examples.

Rational Number	Example
Every natural number is a rational number with a natural number as numerator and 1 as denominator.	$7 = \dfrac{7}{1}$ and $13 = \dfrac{13}{1}$
Every fraction is a rational number with a natural number as numerator and natural number as denominator.	$\dfrac{2}{3}$ and $\dfrac{19}{67}$
The number 0 is a rational number with 0 as numerator and any natural number or its opposite as denominator.	$0 = \dfrac{0}{11}, \dfrac{0}{-11}$
Every integer is a rational number with denominator 1.	$-5 = \dfrac{5}{-1}$ or $\dfrac{-5}{1}$

rational number: A number that can be expressed in the form $\dfrac{a}{b}$, where a and b are integers and $b \neq 0$.

It is important to discuss several facts about rational numbers.

Terminating and Repeating Decimals

Students learn to write fractions as decimals. For example,

$$\frac{7}{10} = 0.7 \text{ and } \frac{5}{11} = 0.454545\ldots$$

Notice that the decimal 0.7 terminates and the decimal 0.454545… repeats.

In general, every rational number can be written as a terminating or repeating decimal, and vice versa.

$$\text{rational numbers} \quad \Leftrightarrow \quad \text{terminating or repeating decimals}$$

Division by 0

A number cannot be divided by 0. This implies that rational numbers cannot have 0 as denominator. To explain why division by 0 is not allowed, consider each division problem below.

$$\frac{7}{100} \quad \frac{7}{10} \quad \frac{7}{0.1} \quad \frac{7}{0.01} \quad \frac{7}{0.001} \quad \frac{7}{0.0001} \quad \frac{7}{0.00001} \quad \frac{7}{0.000001}$$

As you read from left to right, each quotient gets increasingly large. In fact, if you let the denominator get increasingly small, that is, approach 0, then the quotient will become infinitely large. This is why division by 0 is not allowed.

Alternatively, consider $\frac{7}{0}$. Suppose that $\frac{7}{0} = b$, where b is a number.

$$\frac{7}{0} = b$$

$$7 = b \cdot 0 \qquad \leftarrow \text{\textit{Multiplication Property of Equality}}$$

$$7 = 0$$

This is a false conclusion. Therefore, division by 0 is not possible.

Density of Rational Numbers

Between the integers 0 and 1, there are no integers. The inequalities below suggest that there is an infinite number of fractions between 0 and 1.

$$0 < \frac{1}{2} < 1 \qquad 0 < \frac{1}{3} < \frac{1}{2} < 1 \qquad 0 < \frac{1}{4} < \frac{1}{3} < \frac{1}{2} < 1$$

A similar argument can be made for any pair of consecutive integers, for example, between 4 and 5.

$$4 < 4\frac{1}{2} < 5 \qquad 4 < 4\frac{1}{3} < 4\frac{1}{2} < 5 \qquad 4 < 4\frac{1}{4} < 4\frac{1}{3} < 4\frac{1}{2} < 5$$

Between any two rational numbers, there are infinitely many rational numbers. You can say that the set of rational numbers is *dense* in the set of real numbers.

Number of Rational Numbers

There are infinitely many natural numbers. Although the proof is an advanced one, it is possible to show that there are "as many" rational numbers as there are natural numbers. Indeed, there is more that can be said.

Number of Rational Numbers

- There are "as many" fractions as there are natural numbers.

- There are "as many" opposites of fractions as natural numbers.

- There are "as many" rational numbers as there are positive rational numbers.

- There are "as many" rational numbers between 0 and 1 as there are between 0 and 10.

When students learn to add two rational numbers with different denominators, they use rules, number sense skills, and procedures.

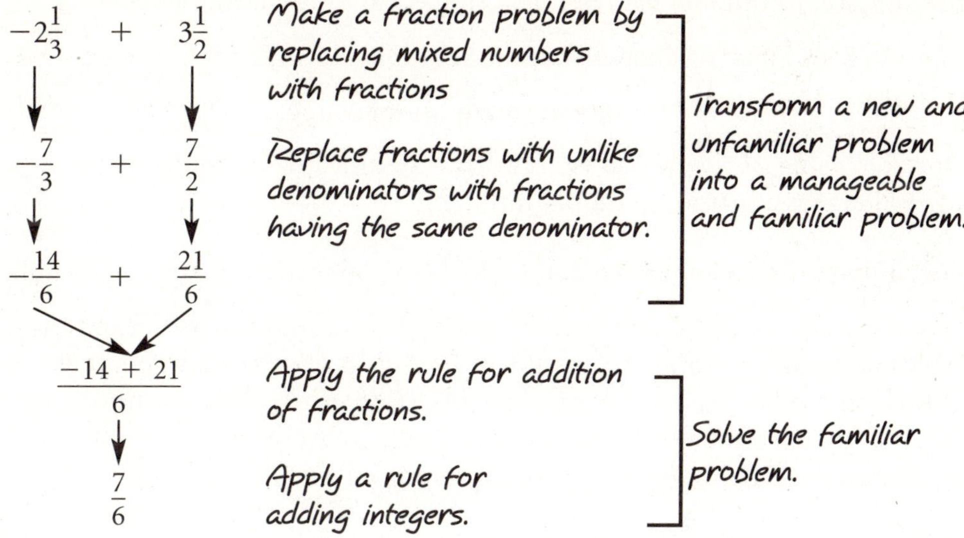

In an introductory algebra course, students continue to reinforce and extend their number sense skills while beginning to reflect on operations, strategies, and abstractions.

Rational Expressions

Just as a polynomial is a generalization of the concept of integer, a **rational expression** is a generalization of the concept of rational number. Since the numerator and denominator of a rational number are integers, you can expect that the numerator and denominator of a rational expression will be polynomials.

$$\text{rational number: } \frac{-7}{13} \quad \begin{array}{l} \leftarrow \\ \leftarrow \end{array} \quad \frac{\text{integer}}{\text{nonzero integer}}$$

$$\text{rational expression: } \frac{x^2 + 5x + 6}{x + 2} \quad \begin{array}{l} \leftarrow \\ \leftarrow \end{array} \quad \frac{\text{polynomial}}{\text{nonzero polynomial}}$$

> ### Definition of a Rational Expression in One Variable
>
> A *rational expression* is a quotient of two polynomials written in fraction form.

rational expression: If P and Q are polynomials and $Q \neq 0$, then an expression of the form $\frac{P}{Q}$ is a rational expression.

Every polynomial is a rational expression whose numerator is a polynomial and whose denominator is 1.

Because a rational expression is defined in terms of polynomials, students need to learn about polynomials before they study rational expressions. An introduction to rational expressions is linked to rational numbers.

A rational number has exactly one value. On the other hand, a rational expression such as $\dfrac{3x + 7}{x - 5}$ has infinitely many values, one value for each allowable value of x you choose. This is a significant difference between a rational expression and a rational number.

Notice that you may find the value of $\dfrac{3x + 7}{x - 5}$ only when x is an allowable value. You may not replace x with 5 because that would cause division by 0. The *domain* of a rational expression is the set of all real numbers that are allowed as replacements for the variable. *Excluded values* are those real numbers that are not allowed as replacements. Any value of the variable that would cause the denominator to equal zero is an excluded value.

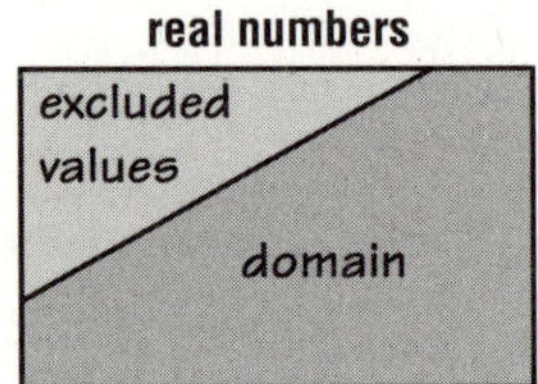

A whiteboard demonstration may help students determine the domain and the excluded values of a rational expression. In this demonstration, students devise a strategy for finding both the domain and the excluded values.

Class: The expression shown here is a rational expression in x.
For what values of x will the expression have a value?
For what value(s) of x will the expression be undefined?
Hint: The denominator may not have 0 as its value.

$$\frac{2x - 1}{x^2 - 13x + 12}$$

Response: If $x^2 - 13x + 12$ cannot be zero, how about solving $x^2 - 13x + 12 = 0$ to find the excluded values? Then all the other real numbers are in the domain.

Class: Good strategy. How do we solve $x^2 - 13x + 12 = 0$?
Response: Factoring might work since the coefficient of x^2 is 1.

$$x^2 - 13x + 12 = (x - 1)(x - 12)$$

Class: Now answer the given questions.
Response: The numbers 1 and 12 are excluded values. Therefore, the domain is all real numbers except 1 and 12.

If students already know how to simplify a rational number, you should prompt them to transfer this knowledge to simplifying a rational expression. The example below illustrates what to do.

simplify a rational number

$$\frac{14}{21}$$

$$= \frac{2 \cdot 7}{3 \cdot 7}$$

$$= \frac{2}{3}$$

simplify a rational expression

$$\frac{x^2 - 13x + 12}{x^2 - 2x + 1}$$

$$= \frac{(x - 1)(x - 12)}{(x - 1)(x - 1)}$$

$$= \frac{x - 12}{x - 1}$$

NAME ____________ CLASS ______ DATE ______

Simplify this rational expression. $\dfrac{x^2 - 5x + 6}{x^2 - 4x + 3}$

First factor the numerator and denominator.

$$\frac{x^2 - 5x + 6}{x^2 - 4x + 3} = \frac{(\quad\quad)(\quad\quad)}{(\quad\quad)(\quad\quad)}$$

Write the numerator and denominator you get when you divide numerator and denominator by common factors.   ____________

The given expression and the expression you just wrote are equal except for what values of x?   ____________

Summarize your work.   ____________________

____

____

After modeling the process of simplifying rational expressions, you can give students independent practice. A worksheet like the one above can be used for this purpose. Make sure students take these three steps.

❶ Factor.

$$\frac{(x - 2)(x - 3)}{(x - 1)(x - 3)}$$

❷ Divide by a common factor.

$$\frac{x - 2}{x - 1}$$

❸ Identify excluded values.

1 and 3

$$\frac{x^2 - 5x + 6}{x^2 - 4x + 3} = \frac{x - 2}{x - 1}, \text{ provided that } x \neq 1 \text{ and } x \neq 3$$

When students summarize their work, they have the opportunity to formulate a method in their own words that they can apply to other simplification problems.

Consider the following abbreviated example.

$$\frac{p^2 + p - 56}{p^2 + 13p + 40} = \frac{(p + 8)(p - 7)}{(p + 8)(p + 5)} = \frac{\cancel{(p + 8)}(p - 7)}{\cancel{(p + 8)}(p + 5)} \quad \leftarrow \text{Factor and divide.}$$
$$= \frac{p - 7}{p + 5}$$

From the factorization of the denominator, $p \neq -8$ and $p \neq -5$.

Again, it is important to give students the opportunity to articulate their thinking immediately after they are guided through the steps for solving a problem. As a result, students learn to formulate a general method and transfer skills used in that method to other similar problems.

Rational Functions

A **rational function** is a function defined by a rational expression. Many applications in everyday life provide examples of rational functions.

> Mr. and Mrs. Jefferson need to drive 300 miles. If they change the speed at which they drive, they will change the amount of time it takes for the trip.

rational function: A function of the form $y = \dfrac{P}{Q}$ or $f(x) = \dfrac{P(x)}{Q(x)}$, where $\dfrac{P}{Q}$ is a rational expression.

Reflecting on the context of this problem, you can make some inferences.

- If their speed is infinite, they will cover the distance in no time at all.
- If their speed is 0 miles per hour, they will not cover the distance at all.
- If their speed is 50 miles per hour, they will complete the trip in 6 hours.
- If their speed is 60 miles per hour, they will complete the trip in 5 hours.

The formula that relates distance, rate, and time is given below. Each version of the formula is solved for a different variable.

$$\text{distance} = \text{rate} \times \text{time} \qquad \text{time} = \frac{\text{distance}}{\text{rate}} \qquad \text{rate} = \frac{\text{distance}}{\text{time}}$$

Next, we can substitute the given distance, 300 miles, in the formula.

$$300 = \text{rate} \times \text{time} \qquad \text{time} = \frac{300}{\text{rate}} \qquad \text{rate} = \frac{300}{\text{time}}$$

Finally, we can make a table of calculations.

rate (speed)	10	20	30	40	50	60	70	80	90	100
time	30	15	10	7.5	6	5	≈4.3	3.75	≈3.3	3

Notice that the faster they drive, the less time it will take to make the trip. That is, as speed increases, time needed for the trip decreases. The illustration at right shows this relationship.

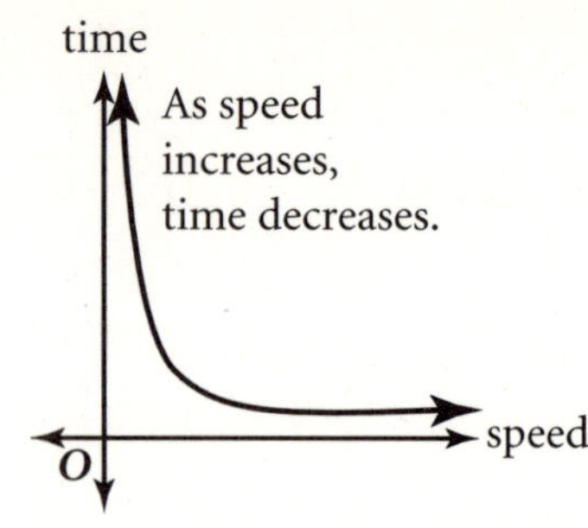

You may say that *y varies inversely* as *x*. This statement can be translated into an equation. For a nonzero constant, *k*, if $xy = k$ or $y = \dfrac{k}{x}$,

then there is an inverse relation between *y* and *x*. In this problem, the fixed distance is 300 miles. If *s* represents speed in miles per hour and *t* represents time in hours, then $s = \dfrac{300}{t}$ and $t = \dfrac{300}{s}$. Each of $s = \dfrac{300}{t}$ and $t = \dfrac{300}{s}$ is a rational function representing the inverse variation. In the former equation, *s* is a function of *t*. In the latter equation, *t* is a function of *s*.

Both formulas are useful for different purposes. For example, to find how long it will take for the trip if they drive at 45 miles per hour, evaluate $t = \dfrac{300}{s}$ with *s* replaced by 45. It will take $6\frac{2}{3}$ hours. But if you want to find how fast they must drive to complete the trip in 4 hours, evaluate $s = \dfrac{300}{t}$ with *t* replaced by 4. They would have to drive 75 miles per hour.

Operations on Rational Expressions

Addition of rational expressions is like addition of fractions. For example, the sum $\dfrac{3x}{x-2} + \dfrac{5}{x-2}$ is simplified as $\dfrac{3x}{x-2} + \dfrac{5}{x-2} = \dfrac{3x+5}{x-2}$, where $x \neq 2$.

You may guide a discussion on addition of rational expressions by comparing the addition problems shown below.

$$\frac{2}{3} + \frac{2}{9} = \frac{2}{3} \cdot \frac{3}{3} + \frac{2}{9} \qquad\qquad \frac{5}{x} + \frac{3}{2x} = \frac{5}{x} \cdot \frac{2}{2} + \frac{3}{2x}$$

$$= \frac{6}{9} + \frac{2}{9} \qquad\qquad = \frac{10}{2x} + \frac{3}{2x}$$

$$= \frac{6+2}{9} \qquad\qquad = \frac{10+3}{2x}$$

$$= \frac{8}{9} \qquad\qquad = \frac{13}{2x}$$

This comparison may help students see the similarity between numerical addition and addition that involves variables. Some teachers might write $\frac{2}{3} + \frac{2}{9}$ on the whiteboard and have a student come to the board to work it out. The teacher might write $\frac{5}{x} + \frac{3}{2x}$ on the board to the right of $\frac{2}{3} + \frac{2}{9}$ and ask students to come to the board to figure out how to pattern their addition on the solution of the numerical problem to the left.

A whiteboard demonstration can also help students understand addition problems in which the denominators have no common factor.

> Class: Let's perform the addition of rational expressions. To add, remember what you learned about addition of fractions.
> $$\frac{3x}{x-2} + \frac{5}{x+3}$$
> Are the denominators the same? Do they have common factors?
> Response: $x - 2$ and $x + 3$ are different and have no common factors.

Teachers might pause and give students the opportunity to see if they can think of what comes next. Some students might immediately make the connection that the next step is to multiply each expression by a clever choice for 1.

$$\frac{x + 3}{x + 3} = 1 \text{ provided } x \neq -3 \text{ and } \frac{x - 2}{x - 2} = 1 \text{ provided } x \neq 2$$

The activity may continue as shown here.

> Class: What common denominator do we need for the addition? What multiplication do we perform?
>
> Response: Use $(x - 2)(x + 3)$.
>
> Multiply the left expression by $\dfrac{(x + 3)}{(x + 3)}$
>
> Multiply the right expression by $\dfrac{(x - 2)}{(x - 2)}$.
>
> $$\frac{3x}{x - 2} \cdot \frac{x + 3}{x + 3} + \frac{5}{x + 3} \cdot \frac{x - 2}{x - 2}$$
>
> Class: What does addition with the same denominator tell you to do?
>
> Response: Write the common denominator on the bottom of a fraction form. Write the sum of the products from the numerators on top.
>
> $$\frac{(3x)(x + 3) + 5(x - 2)}{(x - 2)(x + 3)}$$
>
> Class: Simplify the numerator. Then simplify the resulting expression if possible.

When students complete the work, they should get the following answer.

$$\frac{3x^2 + 14x - 10}{(x + 3)(x - 2)}, x \neq 2 \text{ and } x \neq -3$$

Notice that the sum of rational expressions is defined for all real numbers except 2 and -3.

Recall that when real number b is subtracted from real number a, this can be written as $a - b$ or $a + (-b)$. This is the same first step students take when they subtract one rational expression from another rational expression.

When explaining addition and subtraction of rational expressions, teachers need to be sure that students know the information in the table on the following page.

Addition and Subtraction of Rational Expressions

Adding or subtracting two rational expressions involves a process similar to adding or subtracting two rational numbers.

The following skills related to working with polynomials must be practiced before working with addition or subtraction of rational expressions.

- factoring a polynomial

- finding the opposite of a polynomial

- finding the least common multiple of two polynomials

- multiplying two polynomials

- combining like terms

- Students also need to review simplifying a rational expression, in particular, dividing numerator and denominator by a common factor.

Students who have some practice in adding and subtracting rational numbers may enjoy the puzzle below.

> I am thinking of a number. Then I take the reciprocal of the number and the reciprocal of 1 more than the number. Then I add the reciprocals. If I want the result to be 0, what should the number be?

Let n be the number.

$$\frac{1}{n} + \frac{1}{n + 1} = 0$$

$$\frac{1}{n} \cdot \frac{n + 1}{n + 1} + \frac{1}{n + 1} \cdot \frac{n}{n} = 0$$

$$\frac{2n + 1}{n(n + 1)} = 0$$

Students may pause at this point in the process and observe that a rational expression has value 0 only when the *numerator* equals 0.

$$2n + 1 = 0$$

$$n = -\frac{1}{2}$$

Students can verify that the number is $-\frac{1}{2}$ by using operations on rational numbers.

The reciprocal of $-\frac{1}{2}$ is -2. One more than $-\frac{1}{2}$ is $\frac{1}{2}$. The reciprocal of $\frac{1}{2}$ is 2. The sum of the reciprocals is $-2 + 2 = 0$.

Consider the following problems as enrichment exercises.

I am thinking of a number. I take its reciprocal and take the reciprocal of 2 more than the number. Then I add the reciprocals and their sum equals 0. What is the number?

ANSWER: −1

I am thinking of a number. I take its reciprocal and take the reciprocal of 3 more than the number. Then I add the reciprocals and their sum equals 0. What is the number?

ANSWER: $-\dfrac{3}{2}$

I am thinking of a number. I take its reciprocal and take the reciprocal of 4 more than the number. Then I add the reciprocals and their sum equals 0. What is the number?

ANSWER: −2

I am thinking of a number. I take its reciprocal and take the reciprocal of k more than the number. Then I add the reciprocals and their sum equals 0. In terms of k, what do you think the number will be?

ANSWER: $-\dfrac{k}{2}$

An enrichment activity consisting of these exercises provides students with an opportunity to make a conjecture and verify the conjecture.

Since students often take more interest when they are actively involved in learning mathematics, the following may illustrate how students can work together to simplify $\dfrac{2x-10}{x^2+6x+9} \cdot \dfrac{4x+12}{x^2-25}$.

Class: Let's work together to simplify
$$\frac{2x-10}{x^2+6x+9} \cdot \frac{4x+12}{x^2-25}.$$
Student A: Factor $2x-10$.   Student B: Factor $4x+12$.
Student C: Factor x^2+6x+9.   Student D: Factor x^2-25.

Responses: $2x-10 = 2(x-5)$ $4x+12 = 4(x+3)$
$x^2+6x+9 = (x+3)(x+3)$ $x^2-25 = (x-5)(x+5)$
Write the given product with polynomials replaced with their factorizations.
$$\frac{2(x-5)}{(x+3)(x+3)} \cdot \frac{4(x+3)}{(x-5)(x+5)}$$
Class: Each of you write the product in simplified form. On the whiteboard, I'll write what you get.

answer: $$\frac{8}{(x+3)(x+5)}$$

More Real-World Problems

A complex fraction is a quotient of two fractions. Such fractions occur in real-world problems. Consider Samantha's trip as discussed below.

Average Speed and Direct Variation

Samantha Jones bicycled from home to a shop 24 miles away at an average speed of 8 miles per hour and returned along the same highway at an average speed of 12 miles per hour. Find her average speed for the entire trip.

Let t_1 represent travel time required to make a trip to the shop and let t_2 represent the travel time required to return from the shop.

$$t_1 = \frac{24}{8} \text{ and } t_2 = \frac{24}{12}$$

average speed for the entire trip:

$$\frac{\text{total distance}}{\text{total time}} \rightarrow \frac{24 + 24}{t_1 + t_2} \rightarrow \frac{24 + 24}{\dfrac{24}{8} + \dfrac{24}{12}} = \frac{2(24)}{3 + 2} = \frac{48}{5} = 9\frac{3}{5}$$

Her average speed over the entire trip is $9\frac{3}{5}$ miles per hour.

Work Rate and Direct Variation

The following problem involves the amount of work done in a given amount of time.

> Kevin sat at his desk asking himself how much of his class project he could complete in a certain amount of time. He reasoned as follows.
>
> If the project takes 10 days to complete and:
>
> a)   I work for 3 days, then I will complete $\dfrac{3}{10}$ of the work.
>
> b)   I work for 7 days, then I will complete $\dfrac{7}{10}$ of the work.
>
> c)   I work for d days, then I will complete $\dfrac{d}{10}$ of the work.

Kevin's daily *work rate* is $\dfrac{1}{10}$ and the portion of work, w, done in d days is given by the function $w = \dfrac{d}{10}$. Work rates come into play when two people work together.

> Suppose Kevin and Suzanna work together on the class project. If Kevin works alone, it will take 10 days to complete the project. If Suzanna works alone, it will take 8 days to complete the project. How long will it take if they work together?

The table below shows the work rate for each student. In the table, t represents the time spent together to complete the entire project.

student	time spent on project	time required to complete project	work rate	portion of project completed
Kevin	t	10 hours	$\dfrac{1}{10}$	$\dfrac{t}{10}$
Suzanna	t	8 hours	$\dfrac{1}{8}$	$\dfrac{t}{8}$

The two work rates give the portion of the project each student completes.

Kevin's contribution		Suzanna's contribution		project completed
$\dfrac{t}{10}$	$+$	$\dfrac{t}{8}$	$=$	1

To find t, solve $\dfrac{t}{10} + \dfrac{t}{8} = 1$.

$$\frac{t}{10} + \frac{t}{8} = 1$$

$$80\left(\frac{t}{10} + \frac{t}{8}\right) = 80 \cdot 1$$

$$8t + 10t = 80$$

$$t = \frac{80}{18} \approx 4.4$$

Working together, they can complete the project in about 4.4 days.

Boyle's Law and Inverse Variation

Rational expressions are used in problems involving inverse variation. An inverse variation occurs when the increase of one variable causes the decrease of another variable. For example, Boyle's law states that, at a constant temperature, the volume, V, of a gas varies inversely with the pressure, P, exerted on it. If V varies inversely with P, then there is a positive constant k such that $VP = k$. In other words, $P = \dfrac{k}{V}$.

Consider the following problem.

What effect on pressure results when the volume of the gas is halved?

Let P' represent the new pressure and $\dfrac{1}{2}V$ represent the new volume.

$$P' = \frac{k}{\frac{1}{2}V}$$

$$= \frac{k}{\frac{V}{2}}$$

$$= \frac{k}{1} \cdot \frac{2}{V} \qquad \leftarrow \textit{Multiply k by the reciprocal of } \frac{V}{2}.$$

$$= 2 \cdot \frac{k}{V}$$

$$= 2P$$

When the volume of the gas is halved, the pressure is doubled.

Illumination and Inverse Variation

The illumination I of a point a distance d from a light source varies inversely as the square of d. The variables I and d are related by $I = \dfrac{k}{d^2}$, where k is a nonzero constant.

How does the illumination of point A compare with that of point B, a point that is twice the distance from the source?

If A is d units and B is $2d$ units from the source, $I_A = \dfrac{k}{d^2}$ and $I_B = \dfrac{k}{(2d)^2}$. Simplify $\dfrac{I_B}{I_A}$.

$$\frac{I_B}{I_A} = \frac{\dfrac{k}{(2d)^2}}{\dfrac{k}{d^2}}$$

$$\frac{I_B}{I_A} = \frac{k}{(2d)^2} \cdot \frac{d^2}{k} \qquad \leftarrow \textit{Multiply } \frac{k}{(2d)^2} \textit{ by the reciprocal of } \frac{k}{d^2}.$$

$$\frac{I_B}{I_A} = \frac{1}{4}$$

If distance is doubled, illumination is one-quarter of what it was.

Solving a Rational Equation

A *rational equation* is an equation containing at least one rational expression. The equation $\dfrac{2x + 1}{3} = 6$ is a rational equation because $\dfrac{2x + 1}{3}$ is a rational expression. Notice that both $2x + 1$ and 3 are polynomials. Although the equation is rational, students learned early on to multiply each side of the equation by 3 to begin the solving process.

Contrast the rational equation above with $\dfrac{3}{2x + 1} = 6$. In this equation, the variable is in the denominator rather than the numerator. Students do not learn early on how to approach this equation because its solution depends on multiplying each side of the equation by a variable expression, $2x + 1$, rather than a number like 3.

How is the equation $\dfrac{3}{2x + 1} = 6$ solved? The answer is that students may multiply each side of the equation by $2x + 1$, but they must check any solution(s) they get to make sure that numbers that are not solutions are not part of the answer.

$$\frac{3}{2x + 1} = 6$$

$$(2x + 1) \cdot \frac{3}{2x + 1} = (2x + 1) \cdot 6$$

$$-3 = 12x$$

$$-\frac{1}{4} = x$$

The solution must be checked.

$$\frac{3}{2\left(-\frac{1}{4}\right) + 1} = \frac{3}{-\frac{1}{2} + 1} = \frac{3}{\frac{1}{2}} = 3(2) = 6 \; ✔$$

Extraneous Solutions

Most exercises involving rational equations have exactly the solutions found in the solution process. However, students should be taught that the solution process sometimes produces what is called an **extraneous solution.**

$$\frac{2}{z^2 - 2z} - \frac{1}{z - 2} = 1 \qquad \frac{2 - z}{z(z - 2)} = 1$$

After performing the subtraction on the left side of the equation, students will arrive at the equation at right above. When students solve the equation, they find that $z = 2$ or $z = -1$.

Check: $\dfrac{2}{2^2 - 2(2)} - \dfrac{1}{2 - 2}$ not defined

$$\frac{2}{(-1)^2 - 2(-1)} - \frac{1}{-1 - 2} = \frac{2}{3} + \frac{1}{3} = 1 \; ✔$$

These examples highlight the importance of checking solutions.

extraneous solution: A solution that is obtained but that does not satisfy the original equation.

Taking a Look Back

As seen in this chapter, it is clear that students must learn a variety of skills in order to work with rational expressions. Encourage students to stay connected with and in control of their own learning by writing key notes about rational expressions. These notes will help them organize what they already learned and what they need to learn about rational expressions.

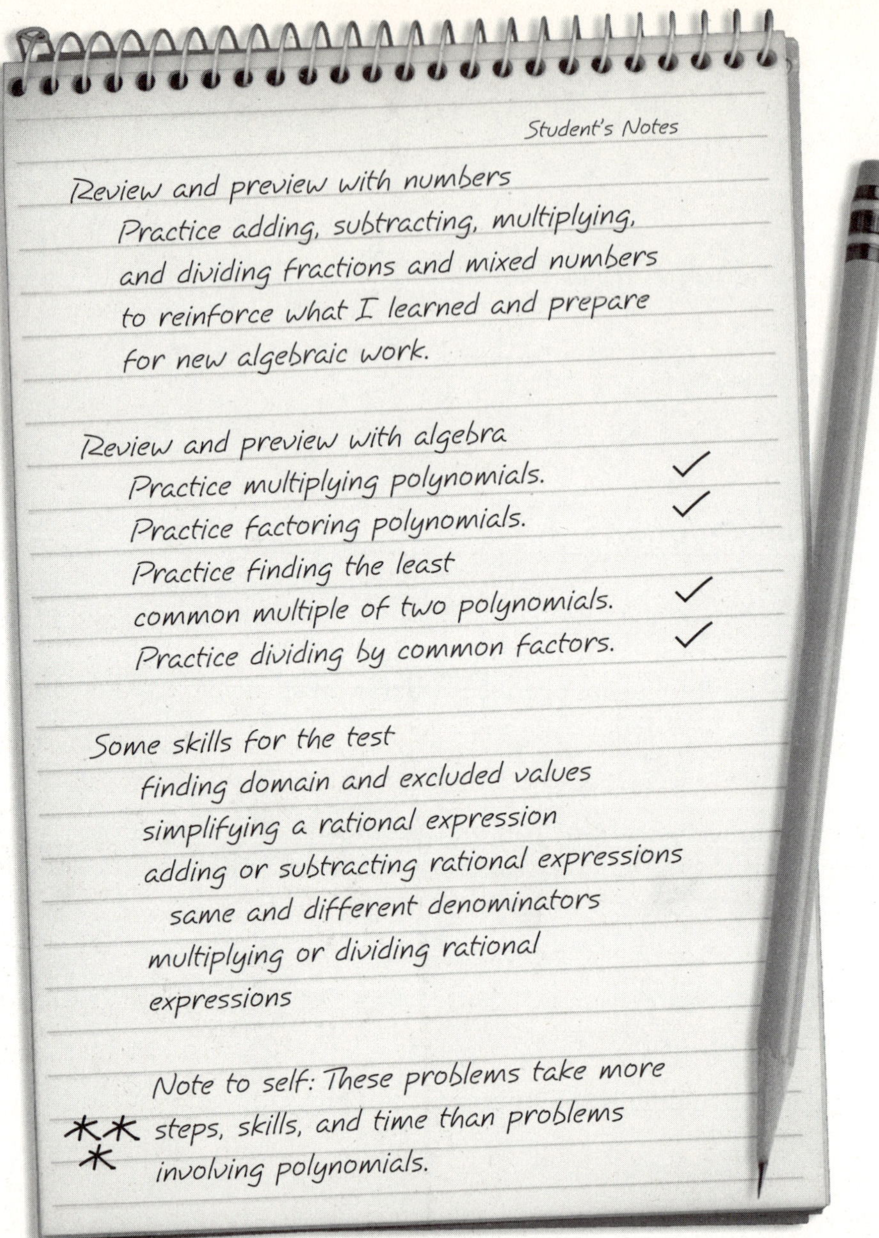

Students can also insert a diagram like the one at right in their journals. This will help them see how far they have come in their study of algebra. Students who know how much they know usually develop a sense of accomplishment.

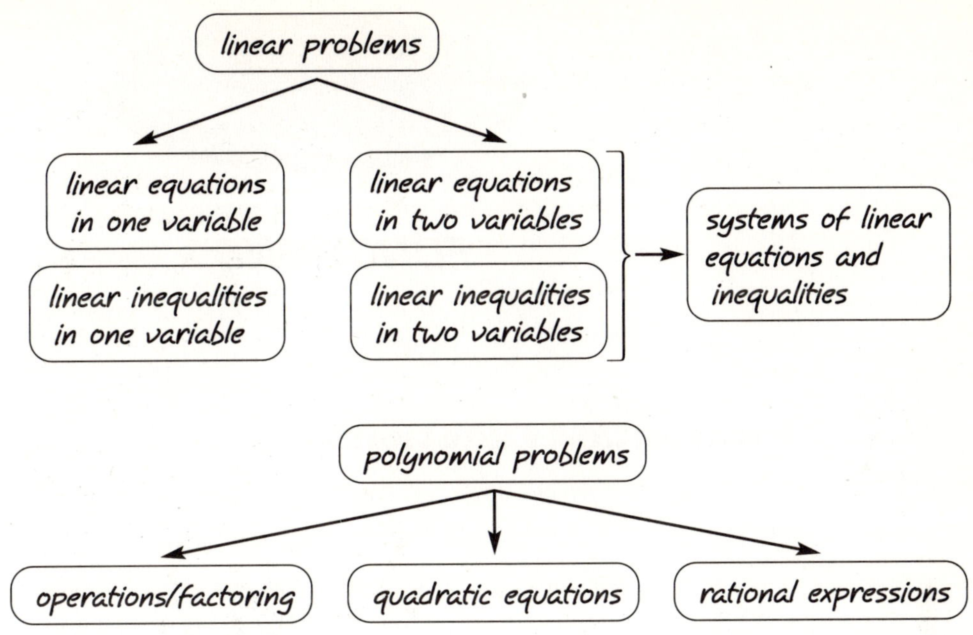

linear problems
linear equations in one variable
linear equations in two variables
systems of linear equations and inequalities
linear inequalities in one variable
linear inequalities in two variables
polynomial problems
operations/factoring
quadratic equations
rational expressions

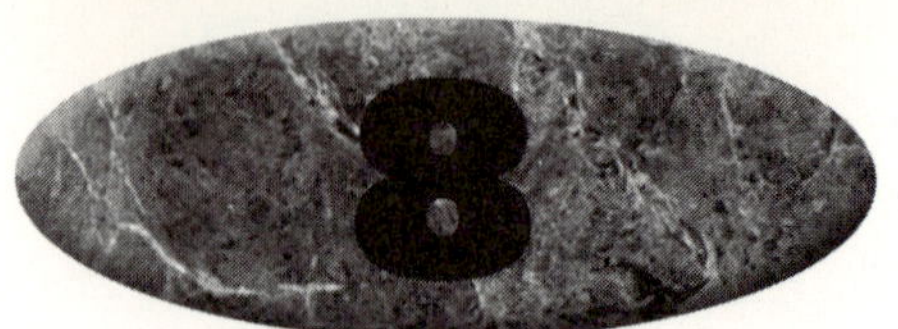

FUNCTIONS

Formulas and Functions

To open our discussion of formulas and functions, consider the following story.

> One day, a homeowner took her trusty tape measure out into the backyard. There she found a square space covered with gravel and rocks. Her intention was to transform that space into a rock garden with various flowering plants arranged here and there. She also wanted to border that region with a homemade wooden fence standing perhaps 18 inches tall. Having measured only one side of the gravel square to be 10 feet long, she concluded that 40 feet of fence material would be needed. Confidently, she drove to the nearest home improvement center and prepared herself for finding just the right materials.

This little story presents a fundamental mathematical concept. For a few moments, reflect on the story.

- If the region is a square and one side measures 10 feet in length, then the total distance around the region is 4 times 10, or 40 feet. This is because the sides of a square are equal.

- There is no other number that gives the total distance around this square. The homeowner can confidently request 40 feet of fencing material without worrying that the quantity needed might be something else.

Mentally, the homeowner applies the formula for the perimeter of a square and knows that the formula gives exactly one perimeter for a given square. Shown below are two formulas that students should learn to recognize and use to find the perimeter of plane figures.

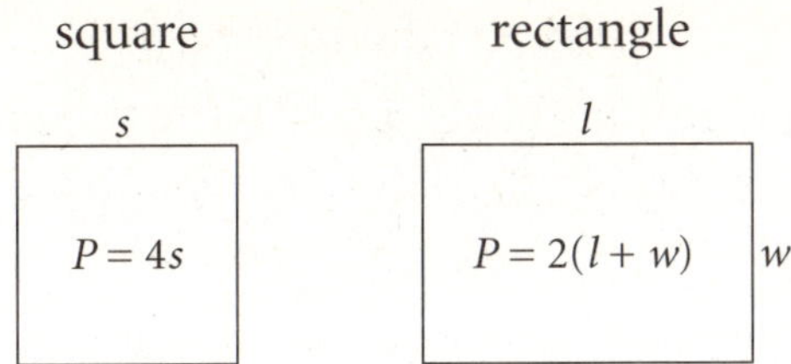

Shown here are several geometry formulas that students should learn to recognize and use to solve area problems related to plane figures.

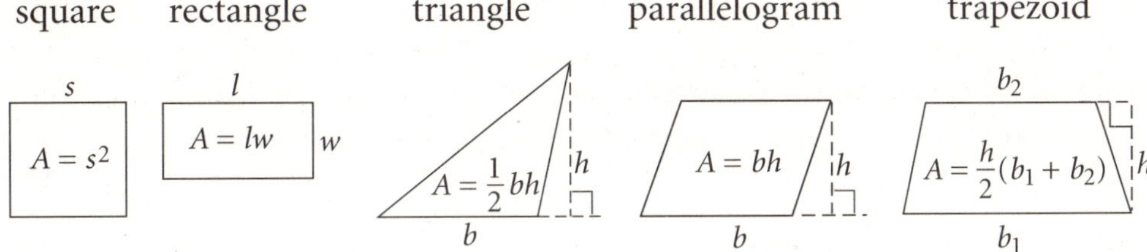

A **formula** is an equation that contains one or more variables whose values you may choose yourself and which gives the value of another variable. For example, in the formula for the perimeter and the area of a rectangle, you choose values for length, l, and width, w. The formula gives the values of P for perimeter and A for area. The variables l and w are called **independent variables** because you choose their values and the variables P and A are called **dependent variables** because their values are determined by the choices you make.

In an introductory algebra course, students learn the *order of operations*.

> ### Order of Operations
>
> ❶ Begin by simplifying expressions within grouping symbols.
>
> ❷ Then simplify expressions involving exponents.
>
> ❸ Carry out multiplication and division in order, left to right.
>
> ❹ Then carry out addition and subtraction in order, left to right.

These rules assure that anybody who evaluates the same formula for the same values of the independent variables will get the same value of the dependent variable.

The illustration below provides a visual way to think of a formula.

independent variable(s) input numbers → formula → dependent variable output number

formula: A literal equation that states a rule for a relationship among quantities.

independent variable: The variable in a function whose value does not depend on the value of the other variable. In x and y notation, usually x is called the independent variable.

dependent variable: The variable in a function whose value depends on the value of the other variable. In x and y notation, usually y is called the dependent variable.

For example, consider doubling a real number.

If you choose any real number, x, and write the product $2x$, then you know from the Closure Property of Multiplication that $2x$ is a unique real number.

$$x \xrightarrow{\ \ doubling\ \ } 2x, \text{ a unique real number}$$

Doubling a real number can be represented by a formula whose input numbers are real numbers and whose output numbers are also real numbers.

The discussion of formulas given above gives rise to the concept of a **function**. A very common way to represent functions is by using equations. For example, each of the equations below represents a function.

$$y = 2x + 3 \qquad y = |x| \qquad y = 2x^2 + 3x - 5$$

All of the functions represented by these equations have one independent variable, x, and one dependent variable, y. You might recognize the functions as a linear function, absolute-value function, and quadratic function, respectively.

function: A pairing between two sets of numbers in which each element of the first set is paired with no more than one element of the second set.

The concept of function is fundamental in mathematics because it provides an important relationship between sets of objects.

> ## Equations and Functions
>
> Suppose that y is given in terms of an expression involving variables. If each replacement of the variables by numbers from given replacement sets gives exactly one value of y, then the equation represents a function.

Input: x	→	Formula: x^2	→	Output
4	→	$(4)^2$	→	16

function

Input: x	→	Formula: $\pm\sqrt{x}$	→	Output
16	→	$\pm\sqrt{16}$	→	4 and -4

not a function

Another way to represent a function is by using a graph. Consider the graph on the following page.

This graph describes the process of filling a tank with water. The horizontal axis represents elapsed time, *t*, in minutes since an intake pipe is opened, and the vertical axis represents the amount of water, *V*, in gallons in the tank after *t* minutes.

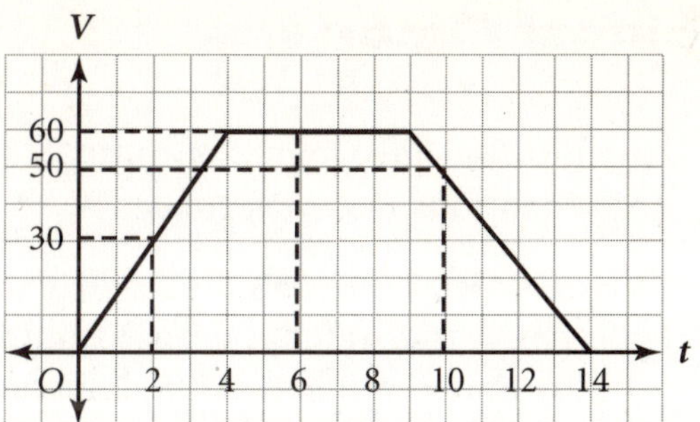

Notice that for each value of *t*, there is exactly one value of *V*.

> After 2 minutes, the tank contains 30 gallons.
> After 6 minutes, the tank contains 60 gallons.
> After 10 minutes, the tank contains 50 gallons.

The graph represents a function but the function is not expressed as an equation.

A function may be represented by a table. Two tables of ordered pairs are shown below. Values of *x* are chosen and values of *y* are determined according to the table. Table A represents a function. Table B does not represent a function.

Table A

x	1	2	3	4	5
y	3	10	17	24	31

Table B

x	1	2	5	4	5
y	3	10	17	24	31

In Table A, for each value of *x*, there is exactly one value of *y*. In Table B, there is a value of *x*, 5, for which there are two different values of *y*, 17 and 31.

$$\text{Table B:} \quad 5 <^{\displaystyle 17}_{\displaystyle 31}$$

This is the reason that Table B does not represent a function.

Graphs, Tables, and Functions

If for each value of *x* there is exactly one value of *y* on the graph, then the graph represents a function.

If for each value of *x* in a table of ordered pairs (*x*, *y*), there is exactly one value of *y* in the table, then the table represents a function.

Linear Functions

The following problem was presented in Chapter 3 as an application of linear equations in two variables.

> A storage tank contains 300 gallons of water. A discharge valve is opened and water is drained out at the rate of 60 gallons per minute. How much water is in the tank after a specified number of minutes?

A whiteboard demonstration can help students discover that this problem involves a function and that the function is linear.

Class: Look at the picture at right. It represents draining a tank. Do you think that volume decreases as time increases?

300 gallons at start

60 gallons per minute

Response: Volume decreases as time increases. For each 1 minute of flow, there is a decrease of 60 gallons of water.

Class: Can you say that volume is a function of elapsed time? Justify your response.

Response: Yes. Volume is a function of time. For each value of t, there is exactly one amount of water in the tank.

Also, we can write an equation to represent V in terms of t.

$$V = -60t + 300$$

This is like a formula. For each value of t, there is just one value of V.

As a follow-up activity, have students represent the function as a graph. The graph is shown at right. Students should notice that the graph is a line segment whose endpoints are (0, 300) and (5, 0). Students should also notice that the graph is *continuous*. That is, the graph has no holes or breaks in it. Allowed values of t are all real numbers between 0 and 5, inclusive. Allowable values of V are all real numbers between 0 and 300, inclusive.

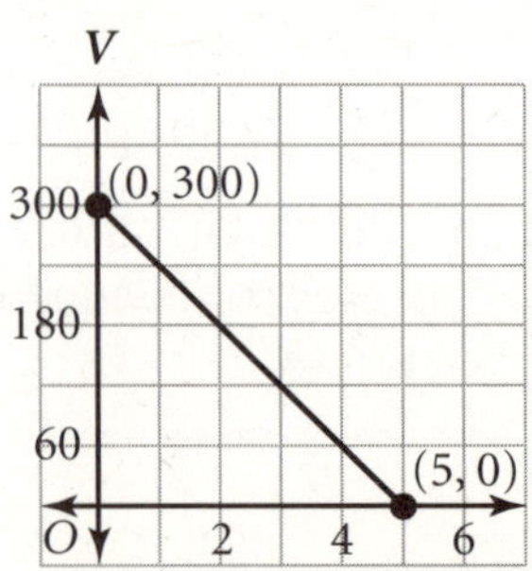

When studying functional relationships, students should become accustomed to describing the behavior of functions. The whiteboard activity below can be used to show students how to describe functional relationships.

Class: I would like individual students to come to the whiteboard and complete each functional relationship with either INCREASES or DECREASES.

As the length of a side of a square increases, its perimeter _______________ .

When the game ends and as time passes, the number of people in the stadium _______________

and the number of people outside the stadium _______________ .

As the number of days a car is rented increases, the rental cost _______________ .

Students can transfer knowledge about functional relationships to other algebra lessons. For example, students may be given the equation $y = 2x + 3$ and be asked to graph it. In Chapter 3, we defined slope of a line as a constant rate of change. That is, if $y = 2x + 3$, then as x increases, so does y. Specifically, for each increase of 1 in x there is an increase of 2 in y. This means that as you move from left to right along the x-axis, the graph should rise to the right.

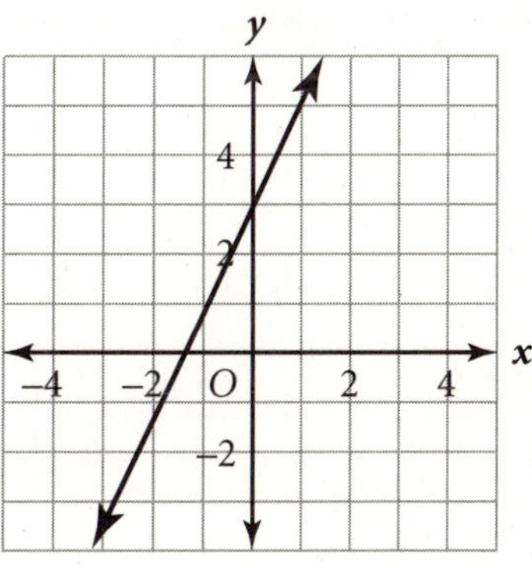

A Step Function

Variations on linear functions often occur in real-world problems. For example, the cost of mailing a letter is a function of the weight of the letter. As weight increases, so does the cost of mailing the letter. However, cost "jumps" with weight, that is, cost does not continuously change as weight changes. To see how this works, consider the following problem.

Suppose that the cost of mailing a first class letter in the United States is $0.33 for the first ounce or fraction thereof and $0.22 for each additional ounce or fraction thereof.

The table below shows the cost, *c*, of mailing a letter with weight *w*.

$0 < w \le 1$	$1 < w \le 2$	$2 < w \le 3$	$3 < w \le 4$
$0.33	$0.55	$0.77	$0.99

The data in the table are shown in the graph at right. Notice that the graph shows a resemblance to the graph of a linear function in that it rises in a predictable way. However, it rises in steps.

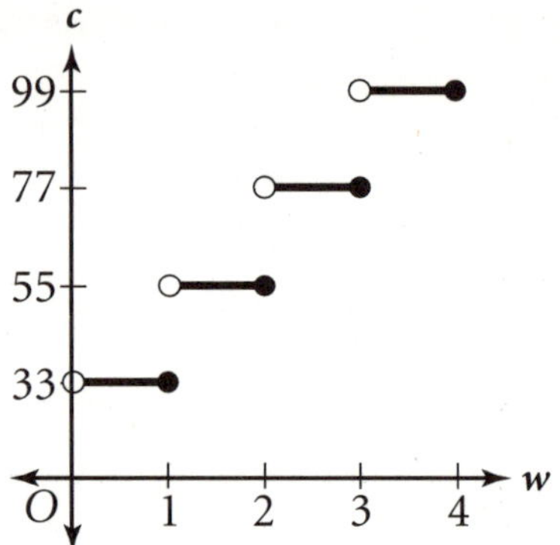

This example indicates that when students read problems, they should take note of whether the situation described in a problem can be represented by a linear function or whether some significant variation of it represents the situation.

Quadratic Functions

The equation $y = x^2$ defines a function because for each value of x you choose, there is exactly one value of y. A close inspection of a table of values can give students an idea of the shape of the graph of $y = x^2$.

x	-3	-2	-1	0	1	2	3
y	9	4	1	0	1	4	9

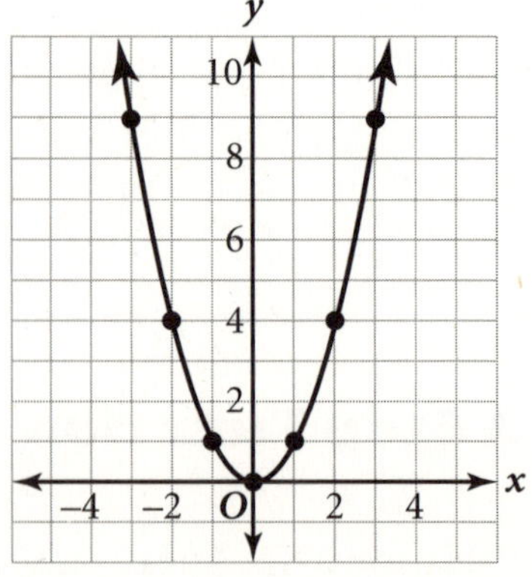

When you graph the ordered pairs in the table, you will get the points shown on the grid at right. If you scan the graph from left to right, you see that this graph decreases, passes through (0, 0), and then increases. When x is negative but increasing, y decreases. When x is positive and increasing, y increases.

Quadratic functions occur in real-world problems. Given below are examples of situations that can be represented by quadratic functions.

- The distance required to stop a moving car varies directly with the square of the car's speed.

- The area of a circle varies directly with the square of its radius.

- The distance traveled by a free-falling object varies directly with the square of elapsed time.

An important observation needs to be made here. A free-falling object does not travel equal distances in equal time periods. This is so because as the object falls, it picks up speed. If no other factors are at work, speed of a free-falling body will increase 32 feet per second each second. The mathematics needed to write an equation relating distance and time is beyond the scope of a beginning algebra course. However, students in beginning algebra do investigate free-falling bodies by using the function below.

$$d = 16t^2 \qquad \begin{cases} t = \text{elapsed time in seconds} \\ d = \text{distance traveled in feet} \end{cases}$$

This is a quadratic function with independent variable t and dependent variable d.

In many situations, a body may have an initial velocity, v_0, in feet per second. The function below represents this situation.

$$d = 16t^2 + v_0 t \qquad \begin{cases} t = \text{elapsed time in seconds} \\ d = \text{distance traveled in feet} \end{cases}$$

Suppose that a ball is dropped from an initial altitude h_0 feet and given initial velocity v_0 feet per second. Using the function $d = 16t^2 + v_0 t$, you can write a function for the altitude $h(t)$ in feet of the ball after falling for t seconds. This function measures the difference between the initial altitude, h_0, and the distance traveled by the ball, d.

$$h(t) = h_0 - (16t^2 + v_0 t) \qquad \begin{cases} t = \text{elapsed time in seconds} \\ d = \text{distance traveled in feet} \end{cases}$$

Students can manipulate the formula to make it look like $h = -16t^2 - v_0 t + h_0$. Typically, this is how the formula is presented in a chapter on quadratic equations and the quadratic formula. Students may be asked how long it will take for the ball to have a specified altitude. This is an opportunity for students to use what they have learned about solving quadratic equations to solve the problem.

> Class: An object has an initial altitude of 600 feet. It is given an initial velocity of 64 feet per second. How long will it take for the object to have an altitude of 300 feet?
>
> Response: Write an equation for altitude in terms of time.
>
> $$h = -16t^2 - 64t + 600$$

The class can continue this exploration by discussing how their knowledge about solving quadratic equations will help them answer the question. The exploration may continue as follows.

> Student A: Write the equation needed to solve the problem.
>
> $$300 = -16t^2 - 64t + 600$$
>
> Student B: Write the equation in standard form.
>
> $$16t^2 + 64t - 300 = 0$$
>
> Student C: Use the quadratic formula to find t.
>
> $$t = \frac{-64 \pm \sqrt{64^2 - 4(16)(-300)}}{2(16)}$$
>
> Class: Use calculators to approximate t.
>
> Response: $t \approx 2.77$ or $t \approx -6.77$
>
> It takes about 2.77 seconds.

The free-falling motion problem has the following pedagogical advantages.

- It involves students in a nontrivial application of mathematics.
- It requires the use of a quadratic function.
- It requires that students apply quadratic equation solving skills.
- It requires students to apply number sense skills as they choose a reasonable solution to answer the question.

Functions and Relations

If you remove the requirement that each x-value has exactly one y-value, the result will be a relation. A **relation** is any relationship between two sets of objects. Students need to know that, although every function is also a relation, not every relation is a function.

relation: A pairing between two sets of numbers

Consider $y \geq \frac{3}{2}x + 2$ and its graph at right.

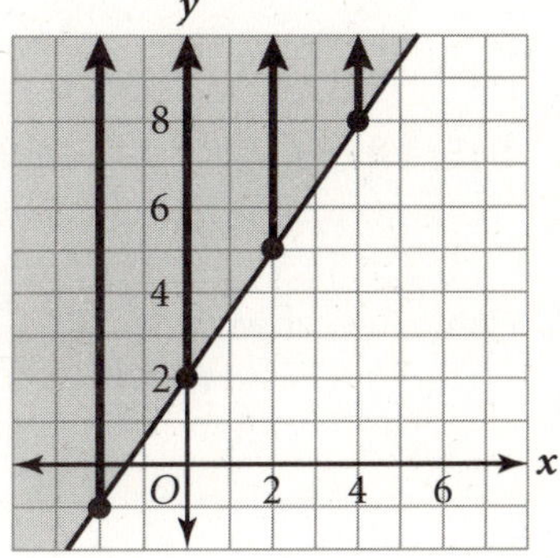

If $x = -2$, then $y \geq -1$.

If $x = 0$, then $y \geq 2$.

If $x = 2$, then $y \geq 5$.

If $x = 4$, then $y \geq 8$.

For each value of x, there are infinitely many values of y. The inequality $y \geq \frac{3}{2}x + 2$ is a linear relation and is not a function.

Students can make the following generalization about linear equations and inequalities.

<table>
<tr><td>

Linear Equations and Inequalities

■ A linear equation containing both variables is a linear function with a line as its graph.

■ A linear inequality containing both variables is a linear relation with a half plane as its graph.

</td></tr>
</table>

The concept of function can be studied in different ways. In our discussion, you have been introduced to functions through the concept of a formula. We have defined a *function* as a correspondence between two sets of objects, the **domain** and the **range,** such that each member of the domain is assigned exactly one member of the range.

domain: The first coordinates in a set of ordered pairs of a relation or function.

There are various kinds of functions. Some types of functions are described in the following table.

range: The difference between the greatest and least values in a data set, or the set of second coordinates in the ordered pairs of a relation or function.

Types of Functions

■ Algebra Functions. You can think of an algebra function as the kind of function that students learn about in algebra. Linear and quadratic functions are examples.

■ Geometry Functions. You can think of a geometry function as a function that assigns something to a geometric object. For example, to each angle in the plane, assign its measure in degrees. As another example, assign to a polygon the number of sides it has.

■ Set Functions. A set function can assign a number to a set of objects. For example, a function can assign to each finite set of objects the number of objects it contains. Another function might assign to a finite set the number of ways you can arrange its members. Such a function might count permutations, ordered arrangements, or combinations, unordered arrangements.

By the time students complete an introductory algebra course, they should be able to:

• identify functions and relations that are represented as a symbolic expression, a table of ordered pairs, or a graph.

• identify the domain and the range of a given function.

• use number sense and function representation to state whether the values of the dependent variable increase or decrease as the values of an independent variable increase.

• write a function involving one independent variable in situations calling for a linear or a quadratic function.

MATHEMATICAL STRUCTURE AND REASONING

The Set of Natural Numbers

Each number in the set of **natural numbers** $\{1, 2, 3, 4, 5, \ldots\}$ can be represented as a finite set of one-to-one correspondences. For example, if a set has 5 elements, we can create exactly 5 one-to-one correspondences between a given set and the set containing 5 elements.

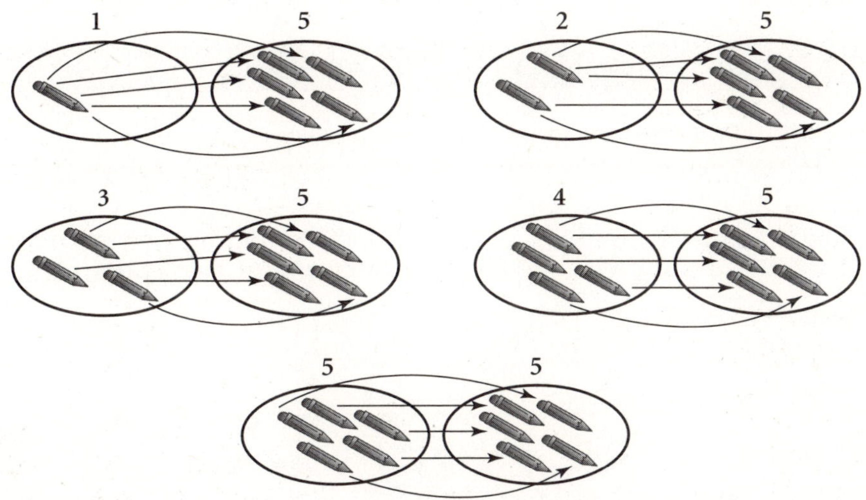

Students should know that addition is a name given to combining sets of objects. For example, if you combine a collection of a pencils with a collection of b pencils, then you will get exactly one new collection of pencils. In general, you can write an abstract statement called the Closure Property of Addition with regard to natural numbers.

Closure Property of Addition

If a and b represent natural numbers, then $a + b$ represents exactly one natural number.

natural numbers:
The numbers 1, 2, 3, 4, 5, …

Consider the sum below.

$$2 + 3$$

The sum represents combining a set of 2 objects with a set of 3 objects. This can be accomplished in two ways.

2 pencils combined with 3 pencils   3 pencils combined with 2 pencils

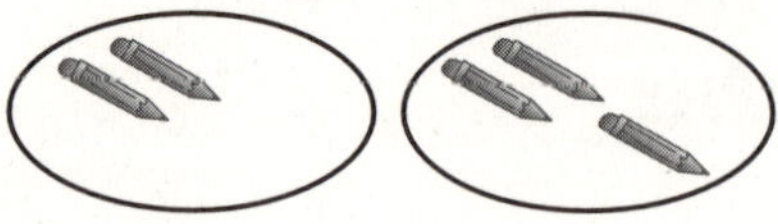 

Using either arrangement gives a sum of 5. Reflection on this simple addition problem suggests a general fact about addition. Whether a collection of a pencils is combined with a collection of b pencils or whether a collection of b pencils is combined with a collection of a pencils, then the same sum results.

This important fact can be formally stated as the Commutative Property of Addition with regard to natural numbers.

Commutative Property of Addition

If a and b represent natural numbers, then $a + b = b + a$.

Addition of natural numbers is extended from adding two numbers to adding three numbers. The diagrams below illustrate the fact that addition is associative.

$$3 + (2 + 3) \qquad = \qquad 3 + 5 \qquad = \qquad 8$$

$$(3 + 2) + 3 \qquad = \qquad 5 + 3 \qquad = \qquad 8$$

Using variables to represent numbers, you can make the following statement about addition of three numbers. If a, b, and c represent natural numbers, then you can state the Associative Property of Addition as written on the top of the next page.

> ### Associative Property of Addition
>
> $a + b + c = a + (b + c) = (a + b) + c$

The previous discussions give you a glimpse into abstract mathematical thinking and can be summarized as follows.

> ### Abstract Mathematical Thinking
>
> - Whenever you add two natural numbers, the sum is exactly one natural number.
>
> - You may add two natural numbers in either order and get the same number for the sum.
>
> - When you add three natural numbers, you may add the first number to the sum of the second and third numbers or add the first two of them and then add the third one and get the same result.

Experimentation and common sense suggest that the general statements above are true without mathematical proof. They seem to be self-evident. In mathematics, a general statement that is accepted without mathematical proof is called an *axiom*. Students in algebra learn early on what are called the properties of addition and multiplication. Some of these properties have been presented as axioms.

In early grades, children learn multiplication of natural numbers as repeated addition. For example, the diagrams below model 3×4 and 4×3.

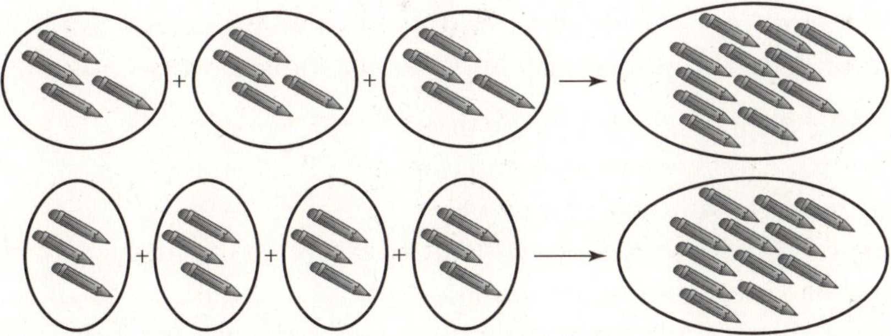

The diagrams illustrate that $3 \times 4 = 12$ and $4 \times 3 = 12$. Three observations you can make about multiplication are that the set of natural numbers follow the Associative and Commutative Properties of Multiplication, and the set is closed under multiplication.

> ### Properties of Multiplication
>
> - When you multiply two natural numbers, you get a unique natural number.
>
> - You may multiply two natural numbers in either order and get the same number for the product.
>
> - To multiply three natural numbers, multiply the first number by the product of the other two numbers or multiply the first two numbers, then multiply by the third number. You will get the same product.

The Set of Integers

whole numbers:
The numbers 0, 1, 2, 3, 4, …

The diagram at right shows a collection of 3 arrows combined with an empty set. Clearly, the sum has as many arrows as 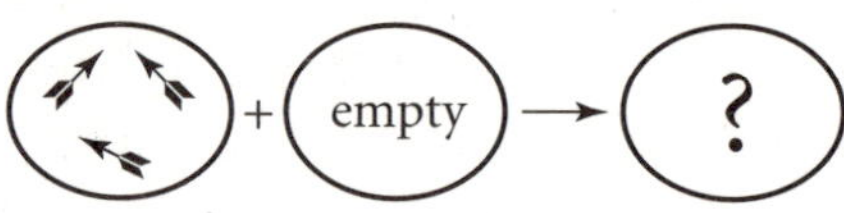 the given set of arrows. The name for the number of members in an empty set is 0. This number, 0, combined with the natural numbers gives the set of **whole numbers.**

$$\text{whole numbers} \qquad \{0, 1, 2, 3, 4, 5, \ldots\}$$

You can call 0 the *additive identity* for the set of whole numbers. In general:

> ### Additive Identity
>
> If a is any whole number, then $a + 0 = 0 + a = a$.

All of the closure, commutative, and associative properties applicable to addition of natural numbers still apply to addition of whole numbers.

Consider the diagram at right. If the first set contains 3 arrows, what must you do in order to have a sum with no arrows?  Clearly, you will have to take away the three arrows in the first set. To indicate placement of 3 arrows in the set, use the symbolic notation $+3$. To indicate the removal of 3 arrows from the set, use the symbolic notation -3. This discussion can serve as an introduction to negative numbers and integers.

A number line can also be used to illustrate integers. The number line at right illustrates the movement from 0 to 3 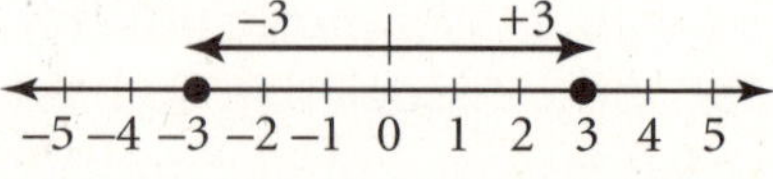 with a right arrow, and the movement from 0 to -3 with a left arrow.

If you combine (add) two equal but opposite movements you will get the two number lines below.

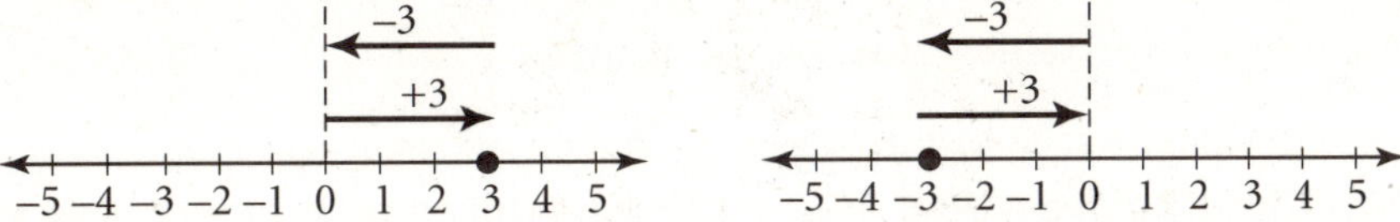

Notice that in both diagrams, you start counting from 0 and then return to 0. Whether you consider sets of arrows or movements on a number line, you will see that the notion of opposite, or additive inverse, is an important one.

The complete set of numbers just created using opposites is called the set of **integers.**

$$\{\ldots, -4, -3, -2, -1, 0, 1, 2, 3, 4, \ldots\}$$

integers: The set of whole numbers and their opposites.

Positive numbers are greater than 0 and *negative numbers* are less than 0. The number 0 is neither positive nor negative. All natural numbers are positive.

The two basic operations on integers are addition and multiplication.

Addition and Multiplication of Integers

- Both addition and multiplication of integers are closed operations. That is, sums and products of integers are unique integers.

- Both addition and multiplication are commutative operations.

- Both addition and multiplication are associative operations.

- The number 0 acts as the additive identity of integers and the number 1 acts as the multiplicative identity of integers.

- Every integer has an opposite, or additive inverse.

- Multiplication distributes over addition:

    If a, b, and c are integers, then $a(b + c) = ab + bc$.

Operations on Integers

The Closure Property of Addition for integers does not tell you how to add two integers. Students usually need substantial practice to learn how to add, subtract, multiply, and divide positive and negative numbers.

Consider the problem below.

> Suppose that Andrea has an outstanding debt of $120. She is unable to pay the debt in full. She is, however, prepared to pay $70 toward paying the debt. If she makes this payment, what will be her debt status?

$$-120 + 70 = -50$$

> She will still have a debt, but it will be reduced to a debt of $50.

The example above shows three important facts.

- Integer addition is useful in solving real-world problems.
- Money and debt may be a useful way to introduce students to integer addition.
- The sum of two integers may be a negative integer.

Addition of integers is perhaps the most difficult operation on integers for students to learn. Once they learn the rules for addition, they will find that the rules for subtraction depend on the rules for addition. A whiteboard demonstration may help students learn how to add two integers.

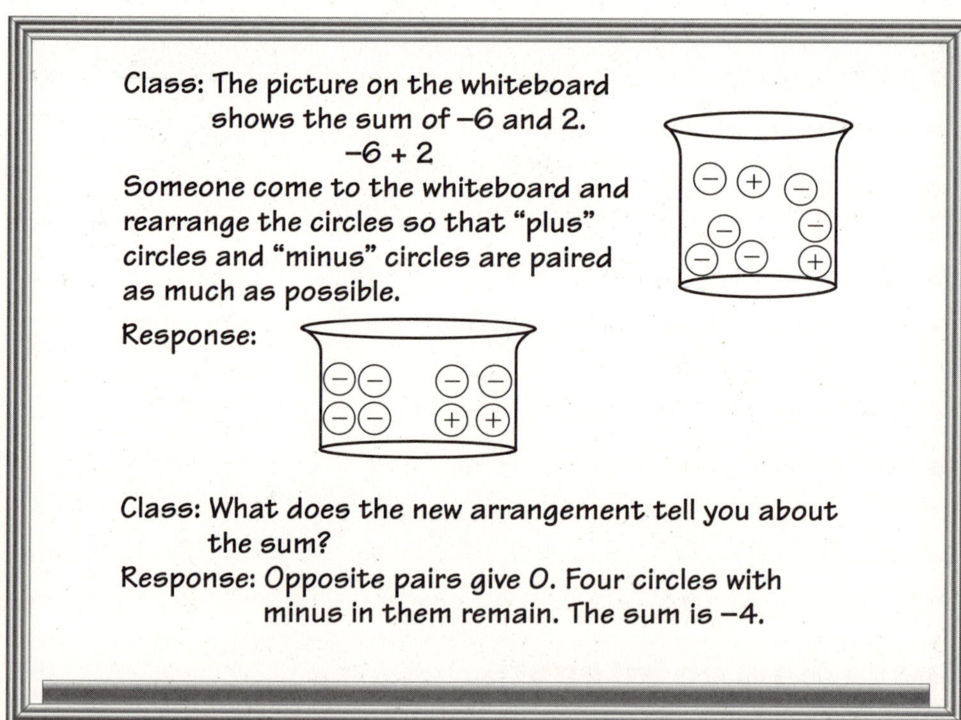

After students explore finding sums by using visual models, they should gradually let operational rules replace these models. It is important to supplement textbook lessons with meaningful discussions of operations on integers so that students can continue their study of solving equations. For example, to solve $x - 6 = -13$, students use the Addition Property of Equality and addition of integers.

$$x - 6 = -13 \qquad \leftarrow \textit{given equation}$$
$$x - 6 + 6 = -13 + 6 \qquad \leftarrow \textit{Addition Property of Equality}$$
$$x = -7 \qquad \leftarrow \textit{adding} -13 \textit{ and } 6$$

Students should be encouraged to reflect on what they have learned. These reflections can help students categorize their knowledge. For example, in the diagram at right, students can see the universe of numbers expand from natural numbers to whole numbers and then to integers.

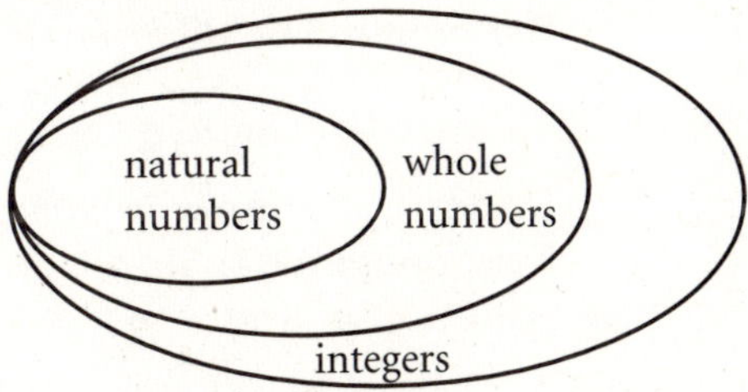

As for solving equations, students progress from solving equations involving whole numbers and positive solutions to solving equations involving either positive or negative numbers.

<table>
<tr><td align="center">simple equation solving</td><td align="center">more advanced equation solving</td></tr>
<tr><td align="center">$x - 6 = 13$</td><td align="center">$x - 6 = -13$</td></tr>
<tr><td align="center">$x - 6 + 6 = 13 + 6$</td><td align="center">$x - 6 + 6 = -13 + 6$</td></tr>
<tr><td align="center">$x = 19$ positive</td><td align="center">$x = -7$ negative</td></tr>
</table>

Rational Numbers

The concept of rational numbers evolves from the concept of fractions. Students learn that a fraction is a quotient like the one at right.

$$\text{fraction:} \ \frac{\text{natural number } a}{\text{natural number } b}$$

Then students learn a generalization of fraction. A rational number is a quotient of two integers.

$$\text{rational number:} \ \frac{\text{integer } a}{\text{nonzero integer } b}$$

Typically, students progress through a learning sequence like this one.

❶ Learn to operate on whole numbers and fractions with positive answers.

❷ Learn to operate on integers with positive or negative answers.

❸ Learn to operate on positive and negative fractional or mixed expressions with positive or negative answers.

The whiteboard activity on the following page shows how the skills identified above come into play in finding a difference such as $4\frac{2}{3} - 7\frac{1}{2}$.

Class: Let's work through the subtraction problem at right. $4\dfrac{2}{3} - 7\dfrac{1}{2}$

Will the difference be positive or negative? Explain.

Response: The difference will be negative because a large number is taken from a smaller one.

Class: The answer is correct. How do we subtract?

Response: Write addition instead of subtraction.

$$4\dfrac{2}{3} - 7\dfrac{1}{2} = 4\dfrac{2}{3} + \left(-7\dfrac{1}{2}\right)$$

Now subtract in the other order.

$$7\dfrac{1}{2} - 4\dfrac{2}{3} = 2\dfrac{5}{6}$$

Class: Are we finished?

Response: No! Attach the negative symbol since the answer is not positive. $-2\dfrac{5}{6} \checkmark$

You may need to assist students with certain prerequisite skills. The diagram below shows skills that are related to subtraction of rational numbers.

subtraction of rational numbers

Review factoring of natural numbers.

Review absolute value.

Review multiplication and addition of integers.

Review subtraction of fractions or mixed numbers.

Every rational number, except 0, has a reciprocal. The product of a number and its reciprocal is 1. For example,

The reciprocal of $\dfrac{1}{3}$ is 3, and $\left(\dfrac{1}{3}\right)(3) = 1$.

The reciprocal of $-\dfrac{1}{4}$ is -4, and $\left(-\dfrac{1}{4}\right)(-4) = 1$.

The reciprocal of 7 is $\dfrac{1}{7}$, and $(7)\left(\dfrac{1}{7}\right) = 1$.

Summarized below are properties of addition and multiplication with regard to rational numbers.

Addition and Multiplication of Rational Numbers

- Both addition and multiplication of rational numbers are closed operations. That is, sums and products of rational numbers are unique rational numbers.

- Both addition and multiplication are commutative operations.

- Both addition and multiplication are associative operations.

- The number 0 acts as the additive identity of rational numbers and 1 acts as the multiplicative identity of rational numbers.

- Every rational number has an opposite, or additive inverse.

- Every rational number except 0 has a reciprocal, or multiplicative inverse.

- Multiplication distributes over addition:

  If a, b, and c are rational numbers, then $a(b + c) = ab + bc$.

Real Numbers

The division calculations below reveal an interesting fact about rational numbers.

$$
\begin{array}{r}
1.375 \text{ terminating} \\
8)\overline{11.000} \\
\underline{8} \\
30 \\
\underline{24} \\
60 \\
\underline{56} \\
40 \\
\underline{40} \\
0
\end{array}
\qquad
\begin{array}{r}
0.583\ldots \text{ repeating} \\
8)\overline{7.000} \\
6\,0 \\
1\,00 \\
\underline{96} \\
40 \\
36 \quad\text{Repeating} \\
40 \quad\text{begins} \\
36 \\
\ldots
\end{array}
$$

Rational Numbers and Decimals

- Every rational number can be written as a terminating or repeating decimal.

- Every terminating or repeating decimal can be written as a rational number.

The following area problem introduces students to square roots of numbers. Remind students that many square roots of numbers are not rational numbers.

Class: We need to find the length of each side of this square. What shall we do?

Response: Guess at x and try multiplying.

$$7 \times 7 = 49 \text{ and } 8 \times 8 = 64$$

The length is somewhere between 7 feet and 8 feet.

Class: Very good! Can you get closer to the actual value of x?

Response: Guess at x and try multiplying again.

$$7.3 \times 7.3 = 53.29 \text{ and } 7.4 \times 7.4 = 54.76$$

Class: Good approximation. The length is between 7.3 feet and 7.4 feet.

irrational numbers: Numbers whose decimal part never terminates or repeats.

The square roots of numbers in this area problem are not rational numbers. Such numbers are called **irrational numbers**. A **real number** is any decimal you can write down. If the decimal terminates or repeats, it is rational. Otherwise, it is irrational. For example, $\sqrt{54}$ is irrational.

$$\sqrt{54} = 7.348469228\ldots$$

real numbers: The set of rational numbers together with the set of irrational numbers.

Other irrational numbers include:

$$\sqrt{3} = 1.732050808\ldots$$

$$\sqrt{7} = 2.645751311\ldots$$

$$\pi = 3.141592654\ldots$$

Many application problems involve solutions that are irrational numbers. In order to work with such solutions, the irrational number is replaced with a rational number that is a close approximation to the irrational number. To explain this strategy to students, consider the whiteboard demonstration on the next page.

Class: We need to find the length of each side of this square. What shall we do?

Response: Solve $x^2 = 54$.

Class: Very good! Now do it.

Response: $x = \sqrt{54}$. Reject $-\sqrt{54}$.

Class: The solution is correct. How useful is it? What if you need to buy that much fencing material to fence a square area? You can't tell home center people to give you $\sqrt{54}$ feet of fencing material.

Response: Approximate $\sqrt{54}$ with a rational number. Buy about 7.3 or 7.4 feet of material for each side.

Deductive Reasoning

Students use **deductive reasoning** when they work an arithmetic problem. They may not articulate the reasoning. However, when students take logically correct steps to arrive at an arithmetic answer, they are reasoning in a deductive manner. An example will make this clear.

Consider $\frac{1}{2} + \frac{1}{3}$.

$$\frac{1}{2} + \frac{1}{3} = \frac{1}{2} \cdot \frac{3}{3} + \frac{1}{3} \cdot \frac{2}{2} \qquad \textit{Multiplicative identity} \quad \frac{3}{3} = 1 \text{ and } \frac{2}{2} = 1$$

$$= \frac{3}{6} + \frac{2}{6} \qquad \textit{definition of multiplication of fractions}$$

$$= \frac{3 + 2}{6} \qquad \textit{definition of addition of fractions}$$

$$= \frac{5}{6} \qquad \textit{addition of natural numbers}$$

deductive reasoning: The process of concluding that something must be true because it is a specific case of a general principle that is known to be true.

In algebra, students use deductive reasoning when they solve equations. The reasoning shown below is an abbreviated deductive argument as found in textbooks.

$$3(x + 5) - 6 = 15$$
$$3(x + 5) = 21$$
$$3x + 15 = 21$$
$$3x = 6$$
$$x = 2$$

This format is compact, practical, and easy to write. Usually students do not write reasons next to the steps taken but should be able to provide them when asked for them.

In the solution below, many conditional statements (*if-then* statements) are chained in a logical fashion and reasons or justifications accompany each statement. The solution to $3(x + 5) - 6 = 15$ looks more like a logical argument.

Consider $3(x + 5) - 6 = 15$ again.

If $3(x + 5) - 6 = 15$, then, by the Addition Property of Equality, $3(x + 5) = 21$.

If $3(x + 5) = 21$, then, by the Distributive Property, $3x + 15 = 21$.

If $3x + 15 = 21$, then, by the Subtraction Property of Equality, $3x = 6$.

If $3x = 6$, then, by the Division Property of Equality, $x = 2$.

Therefore, if $3(x + 5) - 6 = 15$, then $x = 2$.

From this example, you can see that there is a *Transitive Property of Conditional Statements*.

Transitive Property of Conditional Statements

Suppose that p, q, and r represent statements that are either true or false, but not both simultaneously.

If (p, then q) and (If q, then r), then (If p, then r).

The Transitive Property of Conditional Statements is often used when you are given an equation and your goal is to arrive at a solution. The following whiteboard activity provides an opportunity for students to work together, taking turns at constructing a deductive argument. In the activity, the teacher writes an incomplete *if-then* statement and asks various students to complete it. Each statement begins with the conclusion of the previous *if-then* statement.

Class: We will solve the equation $3(x + 5) - 6 = 15$.
        Let's take the first step and give a reason.
If $3(x + 5) - 6 = 15$, then ___________________________.
Response: $3(x + 5) = 21$ by the Add. Prop. of Eq.

Now let's take the next step.
If $3(x + 5) = 21$, then ___________________________.
Response: $3x + 15 = 21$ Distributive Prop.

Let's take another step.
If $3x + 15 = 21$, then ___________________________.
Response: $3x = 6$ by the Sub. Prop. of Eq.

Finally,
If $3x = 6$, then   ___________________________.
Response: $x = 2$ by the Div. Prop. of Eq.

Class: Let's put the argument together in If-Then
        form.
Response: If $3(x + 5) - 6 = 15$, then $x = 2$.

Counterexamples and Conjectures

Counterexamples are useful to show that a statement is not always true.

Consider the assertion below.

The square of every real number is positive.

The following examples seem to prove the given assertion.

$$2^2 = 4 > 0 \qquad\qquad (-3)^2 = 9 > 0$$

$$\left(\frac{2}{5}\right)^2 = \frac{4}{25} > 0 \qquad\qquad \left(-\frac{4}{3}\right)^2 = \frac{16}{9} > 0$$

As tempting as it may be to accept the statement as true, the statement is not true for the number 0. Clearly, $0^2 = 0$, which is not a positive number.

In order for the assertion "The square of every real number is positive" to be true, it must be true for all real numbers. Since the statement is not true for one of them, 0, the assertion cannot be accepted as true. The following modified assertion is always true and it can be proved to be true.

The square of every nonzero real number is positive.

conjecture: A statement that is believed to be true but is not yet proven.

A **conjecture** is a belief that a statement is true and not yet proven to be true. Some conjectures turn out to be false as stated, others turn out to be true as stated, and yet others remain conjectures for a long time.

Consider the sums below.

$$2 + 2 = 4 \qquad 6 + 12 = 18 \qquad 24 + 30 = 54 \qquad 102 + 14 = 116$$
$$8 + 544 = 552 \qquad 48 + 2 = 50 \qquad 64 + 14 = 78 \qquad 76 + 76 = 152$$

What do these sums have in common? What can be said about them and others like them?

The numbers (addends) in all the sums are even numbers and so is each sum. This observation leads to the following conjecture.

Conjecture: If m and n are even numbers, then $m + n$ is an even number.

Proof: If you represent an even number as a multiple of 2, you can write:

$$m = 2k, \text{ for some integer } k \qquad n = 2p, \text{ for some integer } p$$

Therefore:

$$m + n = 2k + 2p$$

Since the Distributive Property applies to addition and multiplication of real numbers, you can write:

$$m + n = 2(k + p)$$

Since addition of integers is closed, $k + p$ is an integer. Thus, $m + n$ is a multiple of 2 and, therefore, an even number.

Using Problem-Solving Strategies

The role of decision making in mathematics can not be underestimated. For example, to solve a simple equation like $2x + 5 = 19$, students must decide what to do in each step as well as choose an overall strategy.

Consider $2x + 5 = 19$. The goal is to find a solution.

Preliminary Step: Read the equation: "Multiply x by 2. Then add 5. The result of these operations is 19."

Overall Strategy: Use various properties of equality.

Step-by-Step Strategies: Subtract 5, then divide by 2.

$$2x + 5 = 19$$
$$2x = 14 \qquad \leftarrow \textit{Subtract 5.}$$
$$x = 7 \qquad \leftarrow \textit{Divide by 2.}$$

The diagram below organizes problem-solving strategies in a sequential order.

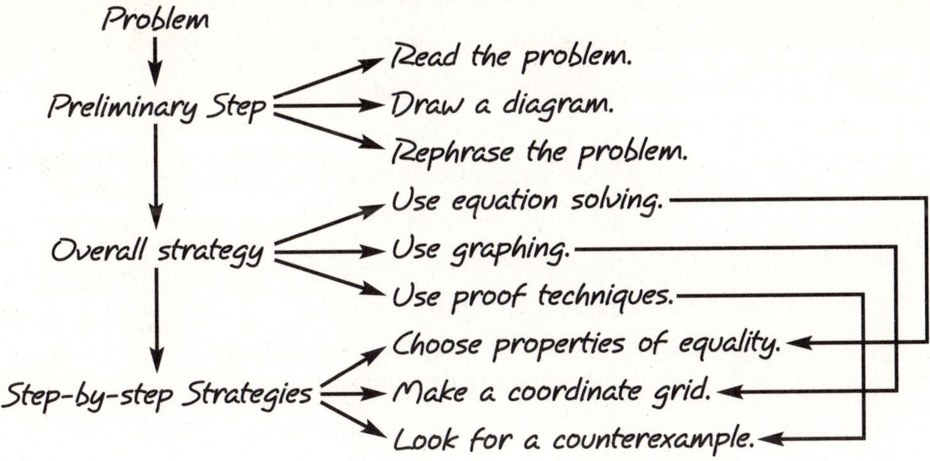

Using Indirect Proof

Deductive reasoning is a process that begins with a **hypothesis,** a statement that is assumed to be true, and ends with a **conclusion.**

Some mathematical hypotheses are easier to prove by using *indirect proof.* According to this type of proof, you assume that the negation of the hypothesis is true and then determine if the conclusion is true.

> For example, the negation of "There are no real numbers such that . . ." is "There is at least one real number such that . . ."

After making the necessary assumptions, you reason logically until you arrive at a contradictory statement. The contradiction tells you to accept the hypothesis and its conclusion as true. A simple example will illustrate the logic of an indirect proof.

Consider the assertion below.

> There is no real number, x, such that $2x - 5 + 3x > 5x$.

> Assume that there is at least one value of x, such that $2x - 5 + 3x > 5x$.

$$2x - 5 + 3x > 5x$$
$$2x + 3x - 5 > 5x \quad \leftarrow \text{\textit{Commutative Property of Addition}}$$
$$(2 + 3)x - 5 > 5x \quad \leftarrow \text{\textit{Distributive Property}}$$
$$5x - 5 > 5x \quad \leftarrow \text{\textit{Addition of numbers}}$$
$$-5 > 0 \quad \leftarrow \text{\textit{Subtraction Property of Inequality}}$$

Since $-5 > 0$ is false, the assumption that there is at least one real number such that $2x - 5 + 3x > 5x$ is false. Therefore, there is no real numbers, x, such that $2x - 5 + 3x > 5x$.

---

hypothesis: In a conditional statement, the portion following the word *if.* In a proof of a conditional statement, the hypothesis is assumed to be true.

conclusion: In a conditional statement, the portion following the word *and.* In a proof of a conditional statement, the conclusion is the part that must be proven to be true.

Consider the assertion below.

There is no nonzero real number, x, such that $x^2 < 0$.

Assume that there is at least one nonzero number, x, such that $x^2 < 0$.

Case 1: Assume that x is positive and that $x^2 < 0$.

$$x > 0 \quad \leftarrow \textit{definition of positive number}$$
$$x \cdot x > 0 \quad \leftarrow \textit{Multiplication Property of Inequality}$$
$$x^2 > 0 \quad \leftarrow \textit{contradiction to assumption}$$

Case 2: Assume that x is negative and that $x^2 < 0$.

$$x < 0 \quad \leftarrow \textit{definition of negative number}$$
$$x \cdot x > 0 \quad \leftarrow \textit{Multiplication Property of Inequality}$$
$$x^2 > 0 \quad \leftarrow \textit{contradiction to assumption}$$

Since there are no positive or negative numbers such that $x^2 < 0$, the given statement must be true.

A Final Look

Mathematical competence depends not only on getting correct answers, but also on developing the ability to reason correctly according to mathematical rules of logic. Consider giving students class projects such as the ones suggested on the following pages.

NAME __________ CLASS _____ DATE _____

Class Project 1

Write a short essay comparing and contrasting reasoning processes used in mathematics with those found in science class.

____

____

____

____

____

____

____

____

____

____

____

____

____

____

____

____

NAME __________ CLASS _____ DATE _____

Class Project 2

Illustrate mathematical problem solving together with mathematical reasoning by writing out a detailed solution to one application problem given in class or home work. Emphasis should be placed on making reasoning and strategies understandable and sensible to the reader. Pictorial representations should be included to enlighten the reader and make clear strategies, as needed.

____

____

____

____

____

____

____

____

____

____

____

NAME ———————— CLASS ———— DATE————

Class Project 3

Write a full solution to the equation
$3(2x + 5) - 7 = 5(x - 3) + 7$.
Include all major steps in logical order with
supporting reasons.
Describe in some detail how previously
learned skills play a role in the solution process.

____

____

____

____

____

____

____

____

____

____

____

____

____

Projects of this nature provide students with the opportunity to:

- demonstrate their knowledge of mathematical procedures.
- demonstrate their knowledge of mathematical concepts.
- demonstrate their knowledge of logic.
- articulate their thinking processes.

Encourage students to place their class projects in their mathematics folder or portfolio. Later, they can look back on their work and reflect on their mathematical maturity. They can also appreciate their ability to read, write, and engage in mathematical discourse with others.

Glossary

absolute value   The absolute value of a real number x is the distance from x to 0 on a number line. The symbol $|x|$ means the absolute value of x. (35)

absolute-value equation   An equation that includes an absolute value. Absolute-value equations have either two solutions or none. (37)

absolute-value inequality   An inequality that includes an absolute value. (37)

Addition Property of Equality   Adding the same amount to two quantities that are equal to one another preserves their equality. (20)

Addition Property of Inequality   If equal amounts are added to the expressions on each side of an inequality, the resulting inequality is true. (29)

Additive Inverse Property   For every real number a, there is exactly one real number $-a$ such that $a + (2a) = 0$ and $-a + a = 0$. (152)

algorithm   A systematic method consisting of a predetermined set of instructions for solving a specific kind of problem. (3)

Associative Property of Addition   For all real numbers a, b, and c, $(a + b) + c = a + (b + c)$. (151)

axis of symmetry   A line that divides the graph of a function into two halves. These halves are reflections of each other. (114)

base of an exponential expression   The number or letter that is raised to an exponent. (81)

boundary line   A line that divides a coordinate plane into two half-planes. (54)

Closure Property   A set of numbers is said to be closed, or to have closure, under a given operation if the result of the operation on any two numbers in the set is also in the set. (77, 149)

coefficient   A number that is multiplied by a variable. (74)

compound inequality   Two inequalities that are combined into one statement by the word and or or. (34)

Commutative Property of Addition   For any numbers a and b, $a + b = b + a$. (150)

conceptual knowledge   The ability to convert concrete to abstract and vice versa. A concept is an abstract generalization, category, or definition with a variety of concrete examples. One goal of mathematics instruction is that students recall a concept that is associated with a specific example. (4)

conjecture   A statement that is believed to be true but is not yet proven. (162)

conclusion   In a conditional statement, the portion following the word *and*. In a proof of a conditional statement, the conclusion is the part that must be proven to be true. (163)

conditional statement   A statement of the form "If *p*, then *q*." (160)

coordinate plane   A plane that is divided into four regions by a horizontal and a vertical number line. (141)

coordinates   An ordered pair of numbers that locates a point in the coordinate plane in relation to the *x*- and *y*-axes. (41)

deductive reasoning   The process of concluding that something must be true because it is a specific case of a general principle that is known to be true. (159)

degree of a polynomial   The degree of a polynomial in one variable is determined by the largest exponent within the polynomial. (73)

dependent variable   The variable in a function whose value depends on the value of the other variable. In *x* and *y* notation, usually *y* is called the dependent variable. (139)

direct variation   If *y* varies directly as *x*, then $y = kx$, where *k* is the constant of variation. (132)

discriminant   For a quadratic equation of the form $ax^2 + bx + c = 0$, where $a \neq 0$, the discriminant is $b^2 - 4ac$. (110)

Distributive Property   For all numbers *a*, *b*, and *c*, $a(b + c) = ab + ac$ and $(b + c)a = ba + ca$, and $(b - c)a = ba - ca$. (85)

Division Property of Equality   Dividing each side of an equation by equal amounts results in an equivalent equation. (22)

Division Property of Inequality   Let *a*, *b*, and *c* be nonzero real numbers. For $c > 0$, if $a > b$, then $\frac{a}{c} > \frac{b}{c}$, and if $a < b$, then $\frac{a}{c} < \frac{b}{c}$. For $c < 0$, if $a < b$, then $\frac{a}{c} > \frac{b}{c}$, and if $a > b$, then $\frac{a}{c} < \frac{b}{c}$. (29)

domain   The first coordinates in a set of ordered pairs of a relation or function. (147)

elimination method   A method used to solve a system of equations in which one variable is eliminated by adding opposites. (63)

equation   Two equivalent algebraic expressions separated by an equal sign. (19)

equivalent equation   An equation having the same solution(s) as the given equation. (22)

extraneous solution   A solution that is obtained but that does not satisfy the original equation. (135)

formula   A literal equation that states a rule for a relationship among quantities. (139)

function   A pairing between two sets of numbers in which each element of the first set is paired with no more than one element of the second set. (140)

greatest common factor   (GCF) The greatest factor that is common to all terms of a polynomial. (82)

hypothesis   In a conditional statement, the portion following the word *if*. In a proof of a conditional statement, the hypothesis is assumed to be true. (163)

independent variable   The variable in a function whose value does not depend on the value of the other variable. In x and y notation, usually x is called the independent variable. (139)

inequality   A statement that two expressions are not equal. Inequalities contain one of the following signs: $<$, $>$, $\leq$, $\geq$, or $\neq$. (27)

integers   The set of whole numbers and their opposites. (153)

intersection (of graphs)   The solution to a system of linear equalities or inequalities consisting of the solutions common to each. (59)

inverse variation   If y varies inversely as x, then $y = k\frac{1}{x}$, where k is the constant of variation, $k \neq 0$ and $x \neq 0$. (126, 133)

irrational numbers   Numbers whose decimal part never terminates or repeats. (158)

linear equation   An equation whose graph is a line. (40)

linear inequality   An inequality whose form is like that of a linear equation with the equal sign replaced by an inequality symbol. (27, 53)

metacognitive skills   Cognitive skills that involve knowing how to organize one's knowledge, how to learn, how to solve problems, and how to correct one's errors in understanding. (10)

monomial   An algebraic expression that is either a constant, a variable, or a product of a constant and one or more variables. (73)

Multiplication Property of Equality
For all real numbers a, b, and c, if $a = b$, then $ac = bc$. (22)

Multiplication Property of Inequality
Let a, b, and c, be nonzero real numbers. For $c > 0$, if $a > b$, then $ac > bc$, and if $a < b$ then $ac < bc$. For $c < 0$, if $a < b$, then $ac > bc$, and if $a > b$ then $ac < bc$. (29)

natural numbers   The numbers 1, 2, 3, 4, 5, . . . (149)

open sentence   A sentence whose truth is not determined until more information is given. (19)

ordered pair   The x- and y-coordinates that give the location of a point in a coordinate plane, indicated by two numbers in parentheses, (x, y). (42)

origin   The point of intersection of the x- and y-axes in the coordinate plane. (41)

parabola   The shape of the graph of a quadratic function. A parabola is defined as the set of all points that are the same distance from a fixed point, called the focus, and a fixed line, called the directrix. (114)

parallel lines   Two nonvertical lines that have the same slope. (66)

perfect square   A number whose square roots are rational numbers. (98)

perpendicular lines   If the slopes of two lines are m and $-\dfrac{1}{m}$, the lines are perpendicular. (66)

point-slope form   The point-slope form for the equation of a line is $y - y_1 = m(x - x_1)$, where the coordinates x_1 and y_1 and are taken from a given point, (x_1, y_1), and m is the slope. (52)

polynomial   The sum or difference of two or more monomials. (72)

principal square root   The positive square root of a number. (98)

procedural knowledge   The ability to follow a predetermined set of procedures. Performing procedures without a conceptual understanding is meaningless learning. (2)

Pythagorean Theorem   If given a right triangle with legs of length a and b and hypotenuse of length c, then $a^2 + b^2 = c^2$. (103)

quadrant   One of the four regions into which a horizontal and vertical line divide a plane. (42)

quadratic equation   An equation of the form $y + ax^2 + bx + c$, where a, b, and c are real numbers and $a \neq 0$. (96)

quadratic formula   For $ax^2 + bx + c = 0$, where $a \neq 0$, $x = \dfrac{-b \pm \sqrt{b^2 - 4ac}}{2a}$. (109)

range   The difference between the greatest and least values in a data set, or the set of second coordinates in the ordered pairs of a relation or function (147)

rate of change   The rate of change of a linear function is equal to the slope of the graph of the function. (48)

rational expression   If P and Q are polynomials and $Q \neq 0$, then an expression of the form $\dfrac{P}{Q}$ is a rational expression. (121)

rational function   A function of the form or , where is a rational expression. (125)

rational number   A number that can be expressed in the form $\dfrac{a}{b}$, where a and b are integers and $b \neq 0$. (118, 155)

real numbers   The set of rational numbers together with the set of irrational numbers. (157)

reciprocal   The number $\dfrac{1}{a}$, is called the reciprocal of a. (156)

relation   A pairing between two sets of numbers. (147)

slope   A measure of the steepness of a line. In the equation $y = mx + b$, m denotes the slope of the line. (49)

slope-intercept form   The slope intercept form for a line with a slope of m and y-intercept of b is $y = mx + b$. (50)

solution to an equation in two variables   The values of x and y that make an equation true. (40)

solution to an inequality in two variables   The set of values of x and y that make up the shaded region representing the solution region to an inequality. (53)

square root   If a is a number greater than or equal to zero, $\sqrt{a}$ represents the positive, or principal square root of a, and $-\sqrt{a}$ represents the negative square root of a. (98)

substitution method   A method used to solve a system of equations in which variables are replaced with known values or algebraic expressions. (60)

Subtraction Property of Equality   Subtracting the same amount from two quantities that are equal to one another preserves their equality. (2, 29)

Subtraction Property of Inequality   If equal amounts are subtracted from the expressions on each side of an inequality, the resulting inequality is true. (29)

system of equations   Two or more equations in two or more variables. (59)

system of inequalities   Two or more inequalities in two or more variables. (67)

term   A number, a variable, or a product or quotient of numbers and variables that is added or subtracted in an algebraic expression. (74)

variable   A letter that is used to represent numbers in an algebraic expression. (40)

vertex   The point where a parabola changes direction. (114)

whole numbers   The numbers 0, 1, 2, 3, 4, . . . (152)

x-axis   The horizontal number line in a coordinate plane. (41)

x-coordinate   A number than indicates distance along the x-axis in a coordinate plane. (41)

y-axis   The vertical number line in a coordinate plane. (41)

y-coordinate   A number that indicates distance along the y-axis in a coordinate plane. (41)

y-intercept   The y-coordinate of the point where a line crosses the y-axis. In the equation $y = mx + b$, b denotes the y-intercept. (50)

Zero Product Property   If a and b are real numbers such that $ab = 0$, then $a = 0$ or $b = 0$. (94, 105)

Index

Graphing
    absolute-value equations and
        inequalities in one variable, 36
    inequalities, 53, 54, 67, 68
    intersection of two lines, 59, 66
    linear equations, 42, 47, 143
    quadratic functions, 113, 115, 116, 144
    step functions, 144
Greatest common factor (GCF), 82
    See also Factoring

Half-plane, 67
Horizontal addition format, 76
Horizontal multiplication format,
    78, 79
Hypothesis, 163

If-then statements, 160
Imaginary numbers, 112
Independent variable, 139
    See also Functions
Indirect proof, 163
Inequalities
    compound, 34
    graphing, 53, 54, 67, 68
    linear inequality in one variable, 27
    restrictions, 68
    solution to an inequality in two
        variables, 53, 54
    strict inequality, 27
    system of inequalities, 67, 68
    weak inequality, 27
Initial state, 48

Input (of a function), 140
Integers, 153
    addition and multiplication, 153
    operations on, 153–155
Intercepts, 42, 43
Inverse Variation, 126, 133
Irrational numbers, 158

Journal, 57, 95, 117, 136

Leading coefficient, 74
Linear equation, 40
    See also Equation
Linear Inequality, 27, 53
    See also Inequality
Lines
    number, 29, 31–32
    parallel, 66
    perpendicular, 66
    slope of, 49

Metacognitive skills, 10
Mixed numbers, 6
Multiplication Property of Equality, 22
Multiplication Property of Inequality, 29

Natural numbers, 149
Number sense, 5

System of inequalities, 67
   See also Inequalities

Transitive Property of Deductive Reasoning, 160
   See also Conditional statement

Variable, 40
Vertex, 114
   See also Parabola
Vertical addition format, 75
Vertical multiplication format, 78, 79

Whole numbers, 152
Whiteboard demonstrations
   adding rational expressions, 127, 128
   anatomy of a polynomial, 73
   determining the domain of a rational expression, 123
   direct and indirect variation, 143
   exploring the discriminant, 110
   factoring a polynomial, 86, 92, 93
   factoring with geometric shapes, 88
   finding a sum by using a visual model, 154
   finding the difference of two mixed numbers, 156
   finding the dimensions of a square, 158, 159
   graphing inequalities (decision chart), 54
   multiplying rational expressions , 131
   simplifying a rational expression, 83
   solving a linear equation, 161
   solving an absolute-value equation, 39
   solving an area problem, 107
   solving a quadratic equation by properties of equality, 102
   solving a quadratic equation by trial-and-error, 97
   vocabulary for polynomials, 74
   writing a linear function for an application problem, 142
   writing a quadratic function for an application problem, 146

x-axis, 41
x-coordinates, 41
x-intercept, 42, 43

y-axis, 41
y-coordinates, 41
y-intercept, 50

Zero-Product Property, 94, 105